ELECTRONICS OF MEASURING SYSTEMS

DESIGN AND MEASUREMENT IN ELECTRONIC ENGINEERING

Series Editors
D. V. Morgan,
Department of Physics, Electronics and Electrical Engineering, University of Wales, Institute of Science and Technology, Cardiff, UK
H. R. Grubin,
Scientific Research Associates Inc, Glastonbury, Connecticut, USA

THYRISTOR DESIGN AND REALIZATION
P. D. Taylor

ELECTRONICS OF MEASURING SYSTEMS
Tran Tien Lang

ELECTRONICS OF MEASURING SYSTEMS

Practical Implementation of Analogue and Digital Techniques

TRAN TIEN LANG
École Supérieure d'Électricité, Gif sur Yvette, France

Translated by the author

English translation edited by

J. McGhee
Department of Electronic and Electrical Engineering,
University of Strathclyde, Glasgow, UK

JOHN WILEY & SONS
Chichester · New York · Brisbane · Toronto · Singapore

Originally published under the title
Électronique des systèmes de Mesures
by Tran Tien Lang

Copyright © 1987 by John Wiley & Sons Ltd.

All rights reserved.

No part of this book may be reproduced by any means, or transmitted, or translated into a machine language without the written permission of the publisher

Library of Congress Cataloging-in-Publication Data:
Tran Tien Lang.
  Electronics of measuring systems.
  (Design and measurement in
electronic engineering)
  Translation of: Electronique des systèmes de mesures.
  Includes index.
  1. Electronic measurements. I. Title. II. Series.
TK7878.T69613   1987   681'.2   86-26786
ISBN 0 471 91157 7

British Library Calaloguing in Publication Data:
Tran Tien Lang
  Electronics of measuring systems.—
  (Design and measurement in
electronic engineering)
  1. Electronic measurements 2. Electronic instruments
  I. Title II. Series III. Electronique des systemes de mesures. *English*
  621.3815'48   TK7878.4

ISBN 0 471 91157 7

Printed and bound in Great Britain

To My Wife

CONTENTS

Series Preface		xv
Preface		xvi

Chapter 1	INTRODUCTION—ABOUT MEASURING SYSTEMS	1
1.1	Fundamental aspects of the contribution of electronics to measurement techniques	1
	1.1.1 Outstanding sensitivity	1
	1.1.2 Very low power consumption	1
	1.1.3 High speed	2
	1.1.4 Great flexibility for remote transmission	2
	1.1.5 High reliability	2
	1.1.6 High versatility in approaching a measuring problem	3
1.2	Measuring systems	3
	1.2.1 Analogue measuring systems	3
	1.2.2 Digital measuring systems	13

PART 1	ELECTRONICS OF ANALOGUE MEASURING SYSTEMS	19

Chapter 2	TRANSDUCER INTERFACING	21
2.1	Passive or modulating transducers	21
	2.1.1 Physical principles	21
	2.1.2 Passive transducer interfacing	26
	2.1.2.1 Signal conditioner for voltage information	27
	2.1.2.2 Signal conditioner for frequency information	33
2.2	Active or self-generating transducers	36
	2.2.1 Physical principles	36
	2.2.2 Active transducer interfacing	36
	2.2.2.1 Voltage-source transducers	37
	2.2.2.2 Current-source transducers	37
2.3	Op amp selection process	38

	2.4	Linearizing	39
		2.4.1 Modification of the transducer circuitry	40
		2.4.2 Processing of the transducer output signal	41
		2.4.2.1 Nonlinearity correction using a logarithmic or exponential converter	42
		2.4.2.2 Nonlinearity correction using an analogue multiplier	42
	2.5	Monolithic transducers	44
		2.5.1 Integrated circuit temperature transducers	44
		2.5.2 Monolithic pressure transducers	45
		2.5.3 Hall-effect integrated circuits	46
Chapter 3		PRACTICAL ASPECTS OF NOISE REDUCTION IN ELECTRONIC MEASURING SYSTEMS	48
	3.1	Definitions	48
	3.2	Examples of noise coupling	49
		3.2.1 Capacitive coupling	49
		3.2.2 Inductive coupling	51
		3.2.3 Conductive coupling	53
	3.3	Grounding connections	53
	3.4	Grounding of cable shields	54
	3.5	Balancing technique	56
		3.5.1 Principle of operation	56
		3.5.2 Common-mode noise voltage	57
		3.5.2.1 Effect of input and source impedance asymmetry	59
		3.5.2.2 Influence of shielded cables	60
	3.6	Bypassing power supplies	61
Chapter 4		AMPLIFIERS	65
	4.1	Instrumentation amplifier	65
		4.1.1 Specifications	66
		4.1.1.1 Gain	66
		4.1.1.2 Offset voltage and input bias currents	66
		4.1.1.3 Frequency response	67
		4.1.1.4 Output impedance	68
		4.1.1.5 Common-mode rejection ratio	68
		4.1.2 Industrial realizations	72
	4.2	Isolation amplifier	76
		4.2.1 Transformer-coupled isolation amplifier	77
		4.2.2 Optically coupled isolation amplifier	77
		4.2.3 Special characteristics of isolation amplifiers	79
		4.2.3.1 Common-mode voltage and isolation voltage	80

		4.2.3.2	Common-mode rejection and isolation-mode rejection	80
		4.2.3.3	Leakage impedance	81
	4.2.4	Applications of isolation amplifiers		81
4.3	Charge amplifier			82
	4.3.1	Necessity for charge amplifier		82
	4.3.2	Principle of operation of a charge amplifier		83
	4.3.3	Imperfections of actual charge amplifiers		84
4.4	Chopper amplifier			85
4.5	Programmable gain amplifiers			88
4.6	Voltage to current converter using instrumentation amplifiers			89

Chapter 5 ANALOGUE SIGNAL PROCESSING FOR MEASUREMENT SIGNALS 92

5.1	Analogue filters		92
	5.1.1	Preliminaries	92
	5.1.2	Passive filters	93
	5.1.3	Active filters	93
	5.1.4	Universal filters	94
	5.1.5	Filters using generalized impedance converters	95
	5.1.6	Rauch cell filters	96
	5.1.7	Switched capacitor filters	98
		5.1.7.1 Principle of operation	98
		5.1.7.2 Basic devices	102
		5.1.7.3 Example of industrial device	109
5.2	Analogue multipliers		111
	5.2.1	Definition—Generalities	111
	5.2.2	Using multipliers in signal processing	114
	5.2.3	Multiplier precision	122
5.3	Logarithmic converters		123
	5.3.1	Definition	123
	5.3.2	Industrial design	124
		5.3.2.1 Stability	125
		5.3.2.2 Temperature compensation	125
		5.3.2.3 Example of monolithic package	129
5.4	A.C./D.C. converters		130
	5.4.1	Absolute-value converters	131
	5.4.2	R.M.S. converter	132
		5.4.2.1 Thermal effect conversion	132
		5.4.2.2 Analogue computing converter	133
	5.4.3	Peak-value converter	135
5.5	Synchronous demodulation		137
5.6	Correlators		139
	5.6.1	Principle	139

		5.6.2	Real-time correlators	140

	5.6.2	Real-time correlators	140
5.6.3	Row correlators	141
5.6.4	Industrial devices	142
5.6.5	Example of application of correlation technique	143
5.6.6	Monolithic correlator	144

5.7 Phase-locked loop (PLL) 145
   5.7.1 Principle of operation 145
      5.7.1.1 The voltage-controlled oscillator 146
      5.7.1.2 The phase comparator 146
      5.7.1.3 The loop filter 147
   5.7.2 Phase-locked loop operation for sinusoidal input 147
   5.7.3 Practical operation of a phase-locked loop 148
      5.7.3.1 Choice of a phase detector 148
      5.7.3.2 Choice of the voltage-controlled oscillator 153
      5.7.3.3 Choice of the loop filter 153
   5.7.4 Loop stability 154
      5.7.4.1 Filter 1: $K(p) = 1/(1 + \tau p)$ 154
      5.7.4.2 Filter 2: $K(p) = (1 + \tau_1 p)/(1 + \tau_2 p)$ 155
      5.7.4.3 Filter 3: $K(p) = (1 + \tau_1 p)/\tau_2 p$ 156
   5.7.5 Locking frequency 156
   5.7.6 Examples of industrial devices 156
   5.7.7 Examples of operation 157
      5.7.7.1 Digital measuring system for phase difference 157
      5.7.7.2 Low disturbance flow-meter 159

PART 2 ELECTRONICS ASSOCIATED WITH DIGITAL MEASURING SYSTEMS 163

Chapter 6 COMPARATORS AND ANALOGUE SWITCHES 165
6.1 Comparators 165
   6.1.2 Basic circuits 165
   6.1.3 Characteristics 167
      6.1.3.1 Accuracy 167
      6.1.3.2 Response time 167
      6.1.3.3 Output voltage levels 168
   6.1.4 Window comparator 169
   6.1.5 Application considerations 169
   6.1.6 Hysteresis comparators 170
      6.1.6.1 Inverting Schmitt trigger 170
      6.1.6.2 Non-inverting Schmitt trigger 171
   6.1.7 Precision Schmitt trigger 172

6.2	Analogue switches	174
	6.2.1 Junction FET switch	175
	6.2.2 MOSFET switches	178
	6.2.3 Diode switch	186
	6.2.4 Bipolar junction transistor switch	187
	6.2.5 Source of error	188
	6.2.5.1 Resistance error in the on state	188
	6.2.5.2 Leakage error	189
	6.2.5.3 High-frequency off isolation	190
	6.2.5.4 Charge coupling and switching time	192

Chapter 7   DIGITAL-TO-ANALOGUE CONVERTERS ANALOGUE-TO-DIGITAL CONVERTERS   193

7.1	Digital-to-analogue converters	193
	7.1.2 Principal types of DACs	193
	7.1.2.1 DAC with weighted resistances	193
	7.1.2.2 DAC with resistance ladders	195
	7.1.2.3 DAC with weighted currents	196
	7.1.2.4 DAC with inverted $R-2R$ ladder	197
	7.1.3 Main DAC specifications	204
	7.1.3.1 Resolution	204
	7.1.3.2 Absolute accuracy	204
	7.1.3.3 Offset error of a unipolar DAC	204
	7.1.3.4 Gain error of a unipolar DAC	205
	7.1.3.5 Linearity	205
	7.1.3.6 Differential non linearity	205
	7.1.3.7 Monotonicity	205
	7.1.3.8 Glitches	206
	7.1.3.9 Settling time	206
	7.1.3.10 Conversion rate	207
	7.1.3.11 Influential factors	207
	7.1.3.12 Notes on high-resolution DACs	207
	7.1.4 Examples of applications	208
	7.1.4.1 Design of a programmable filter	208
	7.1.4.2 Transistor curves tracer system	209
7.2	Analogue-to-digital converters	210
	7.2.1 Parallel ADC. (flash converters)	210
	7.2.2 Successive-approximation ADCs	214
	7.2.2.1 Functional principle	214
	7.2.2.2 Examples of application in instrumentation	218
	7.2.2.3 Microprocessor-compatible ADC	219
	7.2.3 Impulse-counting converters	220
	7.2.3.1 Single-slope converters	220

		7.2.3.2	Voltage-to-frequency converters (VFCs)	223
		7.2.3.3	Dual-slope converters	229
		7.2.3.4	Threefold slope converter	233
		7.2.3.5	Quad-slope converter	234
	7.2.4	Principal specifications of an ADC		237
		7.2.4.1	Hysteresis error	237
		7.2.4.2	Quantization error	237
		7.2.4.3	Missed codes	238
		7.2.4.4	Noise rejection	238
	7.2.5	Examples of applications		240
		7.2.5.1	Structure of a digital filter	240
		7.2.5.2	Structure of a digital storage oscilloscope	242
Chapter 8	SAMPLE-AND-HOLD CIRCUITS—MULTIPLEXERS			247
8.1	Sample-and-hold circuits			247
	8.1.1	Principle of operation		247
	8.1.2	Choice of sampling frequency		247
		8.1.2.1	Sampling theorem	249
		8.1.2.2	Sample-and-hold	251
		8.1.2.3	Relationship between ADC resolution and sampling rate	252
	8.1.3	Principal characteristics of a sample-and-hold amplifier		255
		8.1.3.1	Sample state	255
		8.1.3.2	Hold state	256
		8.1.3.3	Sample-to-hold state	257
		8.1.3.4	Hold-to-sample state	258
	8.1.4	Practical realization		259
		8.1.4.1	Cascade configuration	259
		8.1.4.2	Feedback configuration	260
		8.1.4.3	Sample and hold circuit with integrator	261
	8.1.5	Application of a sample-and-hold in a transient signal-measuring system		262
8.2	Analogue multiplexers			263
	8.2.1	Definition		263
	8.2.2	Channel expansion		264
	8.2.3	Differential multiplexing		266
8.3	Using a multiplexer and sample-and-hold amplifier in a data-acquisition system			267
8.4	Digital multiplexer			268
	8.4.1	Principle of operation		268
	8.4.2	Using digital multiplexer in a data-acquisition system		270

Chapter 9	DATA ACQUISITION AND PROCESSING	272
9.1	Data-acquisition Systems	272
	9.1.1 Building up a data-acquisition system	272
	9.1.1.1 Analogue measurement acquisition	272
	9.1.1.2 Amplification	273
	9.1.1.3 Filtering	273
	9.1.1.4 Sampling rate	274
	9.1.1.5 Conversion	275
	9.1.1.6 Transmission	275
	9.1.2 Industrial realization examples	277
	9.1.2.1 Data-acquisition module	277
	9.1.2.2 Acquisition system controlled by desk computer	277
	9.1.3 Measurement and control system	282
	9.1.4 Measurement stations	285
9.2	Digital processing of data	286
	9.2.1 Digital multiplier-accumulators (MAC)	287
	9.2.2 Digital signal processor	290
9.3	Digital interface for programmable instrumentation	294
	9.3.1 Automatic measurement system	294
	9.3.2 The General-purpose Interface Bus System (GPIB)	295
	9.3.2.1 Description	295
	9.3.2.2 Bus management	296
	9.3.2.4 Simple Examples	298
9.4	Choosing between Analogue and Digital techniques for design of a measurement system	300
	9.4.1 Advantages and drawbacks of Analogue techniques	300
	9.4.1.1 Drawbacks	300
	9.4.1.2 Advantages	302
	9.4.2 Advantages and drawbacks of digital techniques	303
	9.4.2.1 Drawbacks	303
	9.4.2.2 Advantages	303
9.5	Conclusion	305
Appendix	Measurement and Systems of Units of Measurement	307
	A1 Definitions	307
	A2 Systems of units of Measurement	307
	A3 Standards of Measurement	309
Bibliography		313
Index		315

SERIES PREFACE

The crucial role of design in the engineering industry has been increasingly recognized over recent years, with particular emphasis being placed on this aspect of engineering in first and higher degree work as well as continuing education.

This new series of books concentrates on fundamental aspects of design and measurement in electronic engineering and will involve an international authorship. The authors are sought from scientists and engineers who have made a significant contribution in their field. The books in the series will cover a range of topics at research level and are primarily intended for research and development engineers wishing to gain detailed specialist knowledge of design and measurement in a particular area of electronic engineering. It is assumed that, as a starting point, the reader will have a background degree or equivalent qualification in electrical and electronic engineering, physics or mathematics. In the series no attempt will be made to provide preliminary background material but rather the texts will move directly into the design aspects.

Professor D. V. Morgan
Dr H. R. Grubin

PREFACE

Acquisition and processing of information are nowadays the routine work of every technician and scientist, in research laboratories as well as in production plants. Their worktools are undoubtedly dominated by electronics and data processing.

The fast pace of development of these two techniques during the last decades has not always simplified the designer's work, because of the immense possibilities offered by electronics and data processing. In fact, a multitude of functions, always newer and more complex, are realized by monolithic or modular devices. These 'ready to work' products, which may depend on analogue or digital technologies, can exhibit wide variations in both performance and price. Thus, the designer must not only follow recent technological evolution, but also have a full understanding of the numerous and differing technical specifications supplied by the manufacturers, in order to make a discerning choice among the devices that are technically and economically adapted to his application. This book seeks to make a modest contribution towards helping him with this choice. It is not a general electronics book: detailed analysis of some circuits is deliberately omitted. The stress is mainly on the characteristics of commercial products, on criteria of choice, and the care to be taken for their effective practical application.

This book was written principally for practicing engineers. Long mathematical derivations have been avoided on the premise that they are of little interest to these readers. For those readers who may be interested in further details, I have listed numerous references.

Nevertheless because many aspects of electronic functions can be expressed only in mathematical terms, it is desirable that the readers have some mathematical background. For fullest understanding of the subject some familiarity with Laplace Transform and Fourier Transform notation may be needed.

This book would be an opening on the concrete world of today's industrial realizations. It describes several applications including some specialized complete measurement systems that have been designed using advanced electronic circuits. I have avoided producing a circuits cookbook; I have tried to stress physical understanding of basic phenomena as much as possible; specific circuits quickly become obsolete.

This work is the translation of my book *Electronique des systèmes de mesures* published by French Masson Editor.

In undertaking the translation of my book, I tried to bring it up to date. Nevertheless I think that in order for a book on electronics in measuring systems to do full justice to the subject, it would have to be at least an order of magnitude larger than the present volume.

I would like to thank Dr McGhee who reviewed the manuscript of the English Edition for his expert comments and corrections.

I wish to express my gratitude to Professor R. Duperdu for his 'Instrumentation and measurement' course displayed at the Ecole Superieure d'Electricité which was for me a valuable help in the preparation of this book

June 1986   Tran Tien Lang

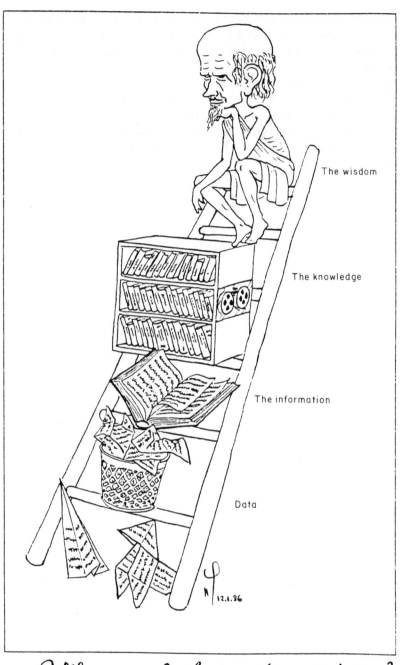

Chapter 1
INTRODUCTION—ABOUT MEASURING SYSTEMS

1.1 FUNDAMENTAL ASPECTS OF THE CONTRIBUTION OF ELECTRONICS TO MEASUREMENT TECHNIQUES

Measurement consists in relating the physical quantity to be determined to another quantity, which is directly accessible to the terminal element of a measuring system. In the past, mechanical and optical means have been used, but nowadays electrical signals are invariably used.

Electrical signals may be obtained, either directly when variation of the physical quantity generates an electrical quantity (thermoelectricity, piezoelectricity, photoelectricity,...), or indirectly when variation of the intermediate quantity is due to some modification of intrinsic properties of the sensitive device (conductivity, permeability, permittivity,...). The measurement of electrical signals has been made easy by electronic means, the advantages of which are now beyond doubt.

1.1.1 Outstanding Sensitivity

The first contribution of electronic processing consists in making the obtained electrical signal more sensitive to measurement.

Example The bridge method compares the behaviour of the sensitive device to fixed elements. If detecting galvanometers can deviate 1 mm for 10 nV across a resistance of a few tens of ohms, electronic techniques permit measurement down to 10^{-14} A. This high sensitivity is sufficient to cope with difficult measurements, particularly in thermometry and calorimetry, which cannot be dealt with by other methods.

1.1.2 Very Low Power Consumption

When the driving power of the signal source is limited the signal is applied to an electronic amplifying system. The use of high-impedance amplifiers, and electronic buffers results in a considerable reduction of input current with low output impedance for driving a deflection instrument. Input impedances of a few tens of megohms are quite common, and gigaohms

($10^9\,\Omega$). may be attained. Input impedance of $10^{12}\,\Omega$ and bias currents of a few nanoamperes are easily attained in FET-input instrumentation amplifiers. Isolation amplifiers, used for instance in biomedical applications, require a practically non-existent driving and biassing power.

1.1.3 High Speed

Some physical phenomena to be measured may show very quick variations. Common deflection type measuring instruments can only give a distorted representation—because their high inertia limits their response to the mean value over some given time interval. Electronics permits several thousand measurements per second. This rate is essential for automatic measurements, so that electronic instruments are suitable for measuring both steady-state and transient conditions. At present digital electronics is invading the measuring field with increasing repercussions. This is often justified by its essential role as for example in the case of the digital phasemeter which is the only effective means of measuring the slippage of a synchronous machine in perturbated operation. It would be unwise to ignore the rise of microprocessors, which can accomplish innumerable intelligent functions, previously unheard of in electrical instrumentation. This is due to their ability to be adapted to varying conditions, their ease of programming and their capability of operating large programs.

1.1.4 Great Flexibility for Remote Transmission

Often measuring points are remote from the measuring instruments or control elements which use the measured value to supervise or command a piece of equipment. Electronic methods offer the only convenient means of information transmission in a majority of industrial applications.

Example Measuring the torque on a motor shaft gives rise to the problem of connecting resistive gauges to the measuring circuits. This problem must be solved by fixing a transmitter to the rotating shaft. The amplitude or frequency variations of an electrical signal are then applied to an antenna, and may be transmitted by centimetric waves.

1.1.5 High Reliability

This is a result of the trend afforded by electronics toward integration of complex functions. We find measurement systems including sensor, converter, amplifier, filter, ... as simple modules, which adapt the ease of use found in classical applications to that of more particular and perhaps more difficult problems. A single printed circuit board may now consist of a complete data acquisition and conversion system, which includes different monolithic devices such as amplifiers, filters, multiplexers, sample-and-hold circuits, converters, etc. Connections are then minimized, and this obviously offers higher reliability.

1.1.6 High Versatility in Approaching a Measuring Problem

This is an important feature of electronic methods. These may be 'direct', as in electrical measurements, but when it comes to obtaining a measurable effect arising from external events (thermal, optical, mechanical, . . .), methods may be 'indirect'. This is often the case when ingenious devices lead to measurement of a quantity which may be of a different energy form and have nothing in common with the original one. So voltage is measured via frequency, or phase via the number of pulses during the time interval between consecutive zero-crossings of the voltages to be compared.

This great diversity of measuring methods, when combined with the numerous possible methods of electronic signal processing (often impossible by other techniques), extends the range and scope of electronic measuring processes.

1.2 MEASURING SYSTEM

Electronic methods used in measurements depend upon the structure of the measuring system, which in turn depends closely on the presentation display and final form of the measurement results.

1.2.1 Analogue Measuring Systems

In analogue methods, the signal, bearing useful information, obtained from the sensor, and transmitted through the system, is related to the measured quantity by a continuous law.

The signal delivered by the sensor undergoes a series of transformations, through a signal conditioner which generally includes:

— A circuit which modifies the structure of signal, in order to make it more adapted to transmission.
— An amplifier which enhances the intensity of signal to make it more perceptible, and less sensitive to interference noise.
— One or more subsequent signal-processing devices. Processing may be merely filtering, which modifies the temporal form of signal (clipping, demodulation, shaping, . . .), in order to eliminate some undesired parasitic actions. Processing may also aim for a result 'corrected and adapted' ready for final use. The processing operations may be very simple, such as filtering, linearization, logarithmic conversion, finding absolute value or r.m.s. value, peak value conversions. They may also be complex, such as detection of signal in noise, correlation between two signals, etc.

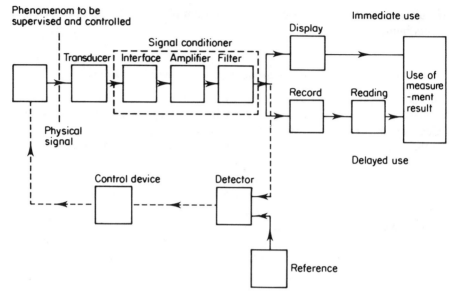

Figure 1.1 Analogue measuring system

Generally speaking, the output value, as received by the data presentation or indicating device, follows the variations of the measured quantity in some continuous proportion.

Example of industrial realization High-speed trains (260 km/h) need a supervision system for the bogie-trucks which are fitted with transverse piezo-resistive accelerometers. When the transverse acceleration of one bogie exceeds 0.8 g for more than 1.5 s, an alarm is sent to the driver's cabin. The measuring signal is recorded for subsequent analysis. The diagram of such a supervision system, as shown by Figure 1.2, includes a sensor which is made of resistive gauges, connected as a measuring bridge acting as a signal conditioner. The accelerometric mass is fitted to one side of a flexible leaf, the other side of which is imbedded as shown in Figure 1.3. Deformation of the leaf is within the limits of bandpass, proportional to acceleration. This measuring system is used by the National French Railways (SNCF = Société Nationale des Chemins de Fer), for supervision of the Paris–Lyon TGV trains (TGV = Très Grande Vitesse).

The physical quantity carrying useful information may have sine-wave form. In this case, the following characteristics may be varied:

— Either the instantaneous amplitude (measured by amplitude modulation)
— Or frequency, or phase, compared with a reference sine wave (measurement by frequency or phase modulation).

Quite often, pulses are used as information carriers. Pulse duration or pulse repetition rate may then be related to the measured quantity by a

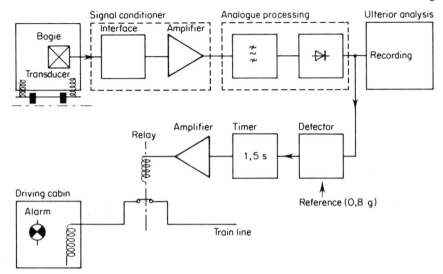

Figure 1.2 Bogie's supervision of high-speed railway train

simple proportional law. Measurement is then discrete, but still remains analogue.

The wide range of configurations allowed by electronics gives a possibility for different signal conditioning components which may result in better discrimination. This is very useful for finer analysis of the measuring method, with the aim of constantly improving the quality of measurement.

Progress in sensor physics, electronics and data processing also contributes to efficiency and economy in organizing the countless measurements which are needed by today's industry. As an example, we mention that supervision and control in a central power station need several thousands measurement points.

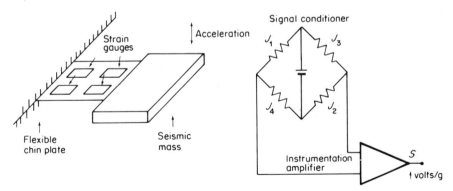

Figure 1.3 Strain gauges are fastened on a flexible chin plate. Transducer interfacing is performed by a bridge with a differential-input instrumentation amplifier

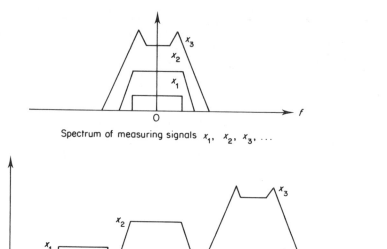

Figure 1.4 Transposition of signal's spectrum

These measurements are geographically scattered in the station and distributed among different physical energy forms such as pressure, temperature, velocity, vibration, liquid level, humidity, etc. In order to avoid cross-talk due to coupling between adjacent channels, and also for economical reasons, several signals are transmitted by a single line. This is known as *multiplexing*. Signal x_i may be represented in the time domain by $x_i(t)$, and in the frequency domain by its Fourier transform $x_i(f)$. Two multiplexing methods are thus distinguished: *time multiplexing*, and *frequency multiplexing*.

Frequency multiplexing

In general, the measuring signals, x_i, are of low-frequency, and although from different sources, occupy the same spectral interval. To avoid cross-talk they must be transposed by a suitable modulation process.

The measuring signals are first filtered to obtain finite spectra, with maximal frequency F_{iM}. An example of a measurement system using the frequency-multiplexing technique is indicated in Figure 1.5. Each filtered signal modulates a sub-carrier at a frequency f_{i0}. Frequencies of the sub-carriers are chosen so that the resulting spectra of modulated signals v_i are distinct, with no overlap. Summing all signals v_i gives then the *multiplex signal*:

$$v = \sum_{i=1}^{n} v_i.$$

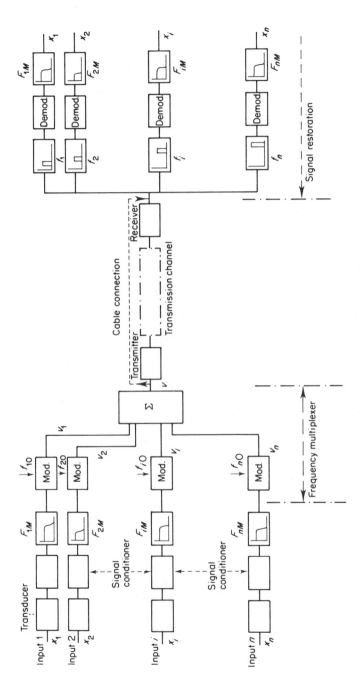

Figure 1.5 Frequency-multiplexing measuring system

If wireless transmission is desired, a second modulation is necessary: this is the modulation of the radio carrier by the multiplex signal v.

To restore the measuring signals, the operations are reversed: band-pass filters isolate the component centered on f_{i0}, and low-pass filters eliminate residual sub-carriers from the demodulated signals.

The frequency shifts needed for frequency multiplexing may be obtained either by amplitude modulation (AM), phase modulation (PM) or frequency modulation (FM).

Notice the following points:

— In amplitude modulation with carrier (AM), the transmitted wave, or modulated signal, is:

$$v(t) = A[1 + mx(t)]p(t)$$

where $p(t) = A \cos 2\pi f_0 t$ is the carrier wave
$x(t)$ is the modulating signal, representing information
m is the modulation index

— In product modulation without carrier, the modulated signal is written

$$v(t) = Kx(t) \cos 2\pi f_0 t$$

— In frequency or phase modulation, the modulated signal is of the form:

$$v(t) = A \sin \phi(t)$$

where $\phi(t)$ is the instantaneous phase.

$v(t)$ is frequency modulated by a signal $x(t)$ when the instantaneous frequency $f(t) = \dfrac{1}{2\pi} \dfrac{d\phi(t)}{dt}$ is a linear function of $x(t)$.

Take

$$f(t) = f_0 + kx(t).$$

Hence

$$\phi(t) = 2\pi \left[f_0 t + k \int_0^t x(\tau)\, d\tau \right] + \phi_0$$

and

$$v(t) = A \sin \left[2\pi f_0 t + 2\pi k_0 \int_0^t x(\tau)\, d\tau + \phi_0 \right]$$

Frequency deviation and modulation index are respectively defined by:

$$f = \sup |kx(t)| \quad \text{and} \quad m = \frac{\Delta f}{F_m}$$

where F_m is the highest frequency of the $x(t)$ spectrum.

Choice of modulation depends generally on the desired signal/noise ratio, and on the precision and bandwidth of the transmission channel. As a general rule, frequency modulation is used when an enhanced signal/noise ratio is required.

Choice of transmission channel is dictated by distance, bandwidth, and external environment conditions. It should be noted that, for interchangeability between nominally similar devices from different suppliers, standardization of the sub-carrier frequencies f_{i0} is required. Two international IRIG standards (InterRange Instrumentation Group), each of which includes 21 channels, are worth mentioning.

(1) Proportional Bandwidth IRIG Standard

In this standard, the maximal deviation of the allowed subcarrier is constant:

$$\frac{\Delta f_{iM}}{f_{i0}} = \text{constant} = 7.5\%$$

Table 1.1 lists the characteristic frequencies of this standard. Note that

Table 1.1 Proportional bandwidth IRIG standard

Channel i	Sub-carrier frequency f_{i0} (Hz)	$f_{i0} - \Delta f_{iM}$ (Hz)	$f_{i0} + \Delta f_{iM}$ (Hz)	Maximal frequency $F_{iM} = \frac{\Delta f_{iM}}{5}$ (Hz)
1	400	370	430	6
2	560	518	602	8
3	730	675	785	11
4	960	888	1 032	14
5	1 300	1 202	1 398	20
6	1 700	1 572	1 828	25
7	2 300	2 127	2 473	35
8	3 000	2 775	3 225	45
9	3 900	3 607	4 193	59
10	5 400	4 995	5 805	81
11	7 350	6 799	7 901	110
12	10 500	9 712	11 288	160
13	14 500	13 412	15 588	220
14	22 000	20 350	23 650	330
15	30 000	27 750	32 250	450
16	40 000	37 000	43 000	600
17	52 000	48 562	56 438	790
18	70 000	64 750	75 250	1 050
19	93 000	86 025	99 975	1 395
20	124 000	114 700	133 300	1 860
21	165 000	152 625	177 375	2 475

the minimum frequency modulation index M_{im} on each channel is 5:

$$M_{im} = \frac{\Delta f_{iM}}{F_{iM}} = 5,$$

where F_{iM} is the maximum frequency of signal x_i. As:

$$\frac{\Delta f_{iM}}{f_{i0}} = 7.5\%,$$

we derive:

$$F_{iM} = \frac{1.5}{100} f_{i0}.$$

thus F_{iM} is proportional to f_{i0}.

(2) Constant Bandwidth IRIG Standard

In this standard f_{i0} frequencies are given by the relationship:

$$f_{i0} = 8(i + 1) \text{ kHz}$$

f_{i0} frequencies lie between 16 kHz and 176 kHz (Table 1.2).

Table 1.2 Constant bandwidth IRIG standard

Series A: 21 channels $\Delta f_{iM} = \pm 2$ kHz $F_{iM} = 400$ Hz		Series B: 10 channels $\Delta f_{iM} = \pm 4$ kHz $F_{iM} = 800$ Hz		Series C: 4 channels $\Delta f_{iM} = \pm 8$ kHz $F_{iM} = 1600$ Hz	
Channel	f_{i0} kHz	Channel	f_{i0} kHz	Channel	f_{i0} kHz
1A	16				
2A	24				
3A	32	3B	32		
4A	40				
5A	48	5B	48		
6A	56				
7A	64	7B	64	7C	64
8A	72				
9A	80	9B	80		
10A	88				
11A	96	11B	96	11C	96
12A	104				
13A	112	13B	112		
14A	120				
15A	128	15B	128	15C	128
16A	136				
17A	144	17B	144		
18A	152				
19A	160	19B	160	19C	160
20A	168				
21A	176	21B	176		

— Series A includes 21 channels: bandwidth Δf_{iM} is ± 2 kHz. Modulation index is $\Delta f_{iM}/F_{iM} = 5$. Hence, maximum frequency F_{iM} on each channel is 400 Hz.
— Series B comprises 10 channels: bandwidth Δf_{iM} is ± 4 kHz, F_{iM} frequency is 800 Hz.
— Series C includes 4 channels: $\Delta f_{iM} = \pm 8$ kHz; $F_{iM} = 1600$ kHz.

In the simultaneous use of these channels, it is obvious that spectral overlapping is avoided. The user should make the choice between either standard, according to the characteristics of the measuring signals. The proportional bandwidth standard is best adapted to signals with a large range of maximum frequencies (between 6 Hz and 4950 Hz). The important criterion for choosing the channels is the desired precision on each channel, depending on the effective spectra of the measuring signals.

Time multiplexing

The entire bandwidth of the transmission channel is dedicated sequentially to each measuring channel, for a defined time interval. Figure 1.7 shows an analogue time-multiplexed measuring system, where we remark:

— Low-pass filters are used, which limit the spectra of the measuring signals to a maximum frequency F_{iM}, in order to permit sampling at a frequency compatible with Shannon's conditions.
— The time-multiplexer is constituted by a series of analogue switches which divert signal from one of the n lines to the output. It is driven by a clock. Measuring signals may be sampled at different frequencies.
— The output signal is a succession of pulses, the amplitudes of which are proportional to the value of the quantity on the line at the sampling constants. It is called 'PAM (Pulse Amplitude Modulation) multiplex signal'.

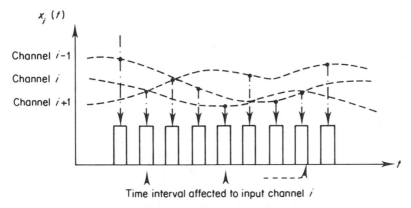

Figure 1.6 In Time multiplexing, the transmission channel is sequentially affected to each input channel

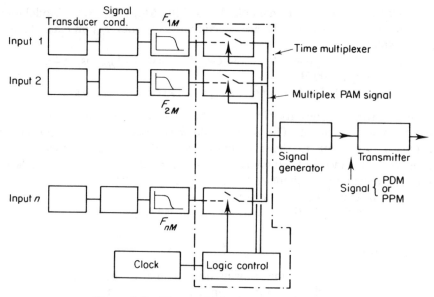

Figure 1.7 Time-multiplexing measuring system

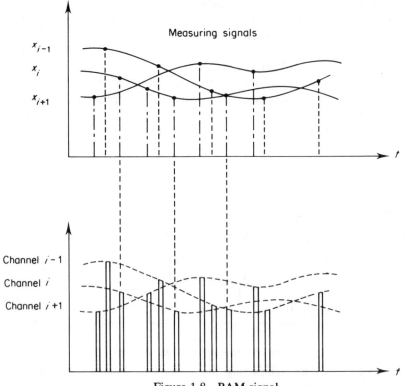

Figure 1.8 PAM signal

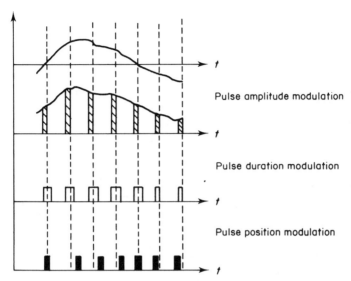

Figure 1.9 Different types of signals to be transmitted after time multiplexing

The PAM signal, where information is carried by the amplitude, may be affected by noise. It can be transformed into:

— PDM (Pulse Duration Modulation) signal, where information is carried by duration of pulse.
— a PPM (Pulse Position Modulation) signal, where information is carried by position of pulse.

Time-multiplexed systems are gradually replacing frequency-multiplexed ones, as they perform better, and in particular lend themselves easily to digitization: the analogue signal is converted to binary words, and the succession of samples forms a PCM (Pulse Code Modulation) message.

In this analogue measuring system, the role of electronics is crucial. As a review of telemetry is not the present aim further information is in the many references quoted in the Bibliography. Thus the present account is limited to the main elements of measuring systems, such as: *signal conditioners, amplifiers,* and the numerous *analogue processing techniques* commonly used in the domain of physical measurement. These elements form the subject of the first part of this book.

1.2.2 Digital Measuring Systems

Progress in digital electronics and data processing is inevitably reflected in the measuring field. The first aim of digital methods is to obtain a number. In digital measuring systems this number appears inside the chain of a

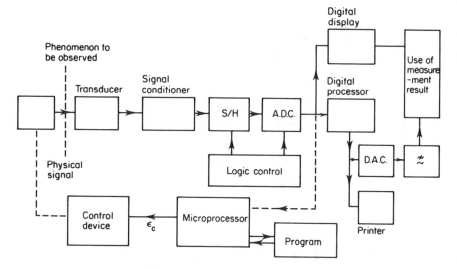

Figure 1.10 Digital measuring system

digital measuring system which is shown in Fig. 1.10. The latter may include a process controller, which may include a microprocessor sequencer.

The signal conditioner is similar to its counterpart in the analogue system. The analogue signal from the conditioner is fed to a *sample-and-hold circuit* (SH), which samples and holds the instantaneous values. Subsequently the hold samples are presented at the input of an *analogue–digital converter* (ADC) for the time duration needed for one conversion. These two modules are driven by a logic circuit which, at chosen times, generates the sampling, holding and conversion commands using logic functions which may be performed by a wired-logic circuit, or a microprocessor software.

The numerical output of ADC may either be processed using 'on-line' or 'off-line' methods by a digital device. These may include a computer to calculate time- or space-averaged values, gradients, RMS values,...). Alternatively they may be stored for subsequent analysis of the evolutions of the physical quantity, or perhaps reconstituted under its analogue form by a *digital–analogue converter* (DAC). Application in process control (temperature control of an oven, quality and production control, etc.) is another important use to which calculated values, may be put.

Example of industrial realization An oven-heating program should present temperature steps, with linear temperature increases and decreases between each step according to the timing diagram in Figure 1.11.

A block diagram of the measuring system is given in Figure 1.12. The temperature sensor is a thermocouple, whose output signal is conditioned using an instrumentation amplifier. The error signal ε is obtained by comparison of the command temperature θ_c, elaborated from program sequence parameters, and the oven temperature θ_o. The latter is derived from the calibration curve of the thermocouple

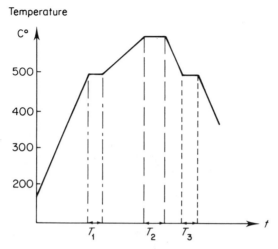

Figure 1.11 Heating program

$\theta_o = f(e)$ which is stored in a 'look-up' table. The sensor's e.m.f. e is converted into a number N by the ADC. The microprocessor determines the oven temperature θ_o from N using the table

The deviation between θ_c and θ_o is used to control the conduction time of a triac, which governs heating of the oven.

This measuring chain has been realized by the Digital Instrumentation and Process Control Laboratory of Angers University (France) for investigating properties of irradiated refractory oxides when cooling to normal state.

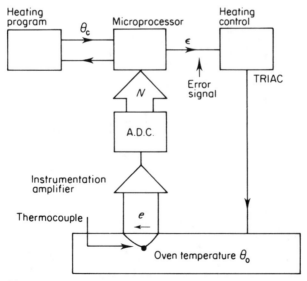

Figure 1.12 Example of a digital-instrumentation and process-control system

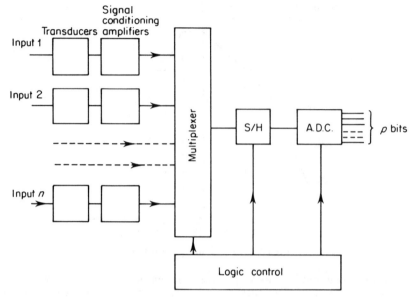

Figure 1.13 First architecture of a data-acquisition system

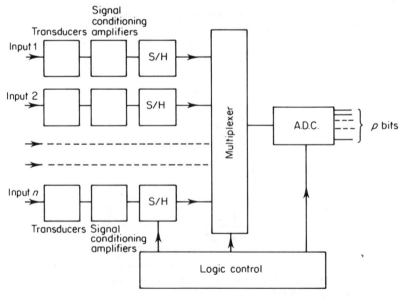

Figure 1.14 Second architecture of a data acquisition system

In most industrial applications, several quantities have to be acquired. A data acquisition system is then made up through time-multiplexing, in a similar manner as seen for the analogue measuring system. This multi-channel acquisition system may be designed in two different configurations, as seen in Figures 1.13 and 1.14.

The *multiplexer,* which is driven by the command logic, promotes an economy of circuits when the ADC and the sample-and-hold circuit are common to all channels (first structure). In some applications, using one *programmable gain amplifier* after the multiplexer allows the removal of the signal-conditioning amplifiers. Notice that, in the first structure, the inputs are successively sampled. However when synchronous data on several channels have to be measured, the second structure, with one sample-and-hold device on each input may be a solution to the problem.

Elementary components of data-acquisition systems now exist as monolithic modules. The object of the second part of this book is to present them with a description of their functional principles, in order to clarify the manufacturers' technical specifications. The ultimate objective is to assist the designer in distinguishing among the many modules available from different manufacturers, so that a proper choice for a particular application may be made. Only the most common methods and principles will be discussed, to avoid the possibility of cataloguing.

PART 1

Electronics of Analogue Measuring Systems

'Share one hub; Adapt the nothing therein to the purpose in hand, and you will have the use of the cart.
Knead clay in order to make a vessel; Adapt the nothing therein to the purpose in hand, and you will have the use of the vessel.
Cut out doors and windows in order to make a room; Adapt the nothing therein to the purpose in hand, and you will have the use of the room.
Thus what we gain is something yet it is by virtue of nothing that this can be put to use.'

Thirty sayings—Taote Ching—Lao Tzu

Part 1

Electronics of Analogue Measuring Systems

Chapter 2
TRANSDUCER INTERFACING

The input quantity for most instrumentation systems is non-electrical. In order to use electrical techniques for measurement the non-electrical quantity has to be converted into an electrical signal by a device called a transducer.

Transducers may be classified according to the electrical principles involved: *passive or modulating transducers* require external power to furnish a voltage or current signal; *active or self-generating transducers* do not require external power as all the electrical energy within the output signal is derived from the physical input.

2.1 PASSIVE OR MODULATING TRANSDUCERS

2.1.1 Physical Principles

These transducers act like electrical impedances. The measurand may produce a variation in geometrical parameters (volume, surface, length, ...) or a change of electrical property (resistivity, permittivity, permeability, ...).

The variation in geometrical parameters is usually due to an applied force, a displacement, a position, vibration, velocity, thickness, ... The variation in electrical properties is usually due to humidity, temperature, pressure, force, torque, light radiation. Tables 2.1 and 2.2 show respectively these two categories of passive transducers.

Example of industrial realizations The concept of converting an applied force into a displacement is basic to many types of transducers. The movable element used to convert the applied force into a displacement is called the force summing device; it may be a diaphragm, a Bourdon tube, a ferromagnetic core, a mass Cantilever, a pivot torque, etc.

Figure 2.1 provides an example of a pressure transducer using a diaphragm as a force summing member. Any variation of the spacing of the parallel plates produces a corresponding variation in the capacitance. This is a direct application of the capacitive transducer principle. When a force is

Table 2.1 Passive transducers using the principle of the variation in geometrical parameters

Physical principles	Nature of device
Resistive Transducers $$R = \rho \frac{L}{S}$$ ρ: resistivity L: length S: surface *Applications*: – displacements – position	$R_1 = R_L \frac{1}{L}$ $R_a = \frac{a}{a_M} R_M$
Capacitive transducers $\varepsilon_0 = 8.85 \, 10^{-12}$ MKSA ε_r = relative permittivity of the dielectric *Applications*: – displacement – vibration – position – pressure – level – sound	*Variation of d* $C = \varepsilon_0 \varepsilon_r \frac{S}{d}$ *Variation of l* $C = \frac{2\pi \varepsilon_0 l}{\ell_n \frac{r_2}{r_1}}$ *Variation of α* $C = \frac{\varepsilon_0 \pi r^2}{360 d} a$ d: distance between the two plates
Inductive Transducer $$L = \mu_0 \frac{N^2 S}{e}$$ L = inductance $\mu_0 = 4\pi 10^{-7}$ S = surface N = Winding number of turns e = air gap	Movable armature / Core / Coil winding

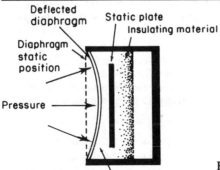

Figure 2.1 Pressure transducer using a diaphragm as force summing member

applied to the diaphragm which operates as one plate of a capacitor, the distance (d) between the diaphragm and the static plate is changed. The resulting change in capacitance may be converted into an electrical voltage or frequency by suitable signal conditioning which will be discussed in the following paragraphs.

In the inductive transducer the measurement of force is performed by the change in the inductance of a coil. This principle of operation is exploited in the manufacture of many excellent displacement transducers.

The linear variable differential transformer (LVDT) is an example of an

Table 2.2 Passive transducers using the principle of variation in electrical properties

Electrical property	Principle of operation	Measurand
Resistivity (ρ)	Resistance strain gauge: Resistivity of a wire or semiconductor is changed by elongation or compression due to externally applied stress. $$\frac{\Delta \rho}{\rho} = C \frac{\Delta V}{V} \begin{cases} V = \text{volume of the wire} \\ C = \text{Bridgman constant} \end{cases}$$	Force, torque, displacement
	Thermistors: These transducers are semiconductor devices that behave as resistors with a high, usually negative, temperature coefficient of resistance.	Temperature
	Photoresistors: Resistivity is a function of the incident electromagnetic radiation.	Light radiation
	Resistance hygrometer: Resistance of a conductive strip changes with moisture content.	
	Resistance thermometer: $\rho = \rho_0(1 + \alpha T°)$	

Type	ρ_0 to 0°C ($\mu\Omega$ cm)	$\alpha \times 10^{-3}$ 0–100°C	Temperature range (°C)
Copper	1.56	4.25	−190–150
Nickel	6.30	5.6	−60–180
Platinum	9.81	3.92	−250–1100
Iridium	9	4.98	−269–27

Permittivity (ε)	*Dielectric gauge*: Variation in capacitance by changes in the dielectric Permittivity is a function of the humidity of the dielectric	Liquid level, thickness Relative humidity

Table 2.2—(continued)

Electrical property	Principle of operation	Measurand
Permeability (μ)	Magnetostriction gauge: magnetic properties are varied by pressure and stress	Force, pressure
	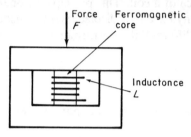	
	$\dfrac{\Delta \mu}{\mu} = \dfrac{\Delta R}{R} = \dfrac{\Delta L}{L} = K\sigma$ R: reluctance $\sigma = \dfrac{F}{S}$ (S: core surface) K: proportionality coefficient	
	The permeability of certain alloys is a function of low temperature. (for example, Vicalloy)	Low temperature (1 K to 250 K)

inductive displacement transducer. It consists of a primary winding and two secondary windings (Figure 2.2). The secondaries have an equal number of turns but they are connected in opposition so that the e.m.f.s induced in the coils oppose each other. The force summing device is a ferromagnetic core; its position determines the flux linkage between the a.c. excited primary winding and each of the two secondary windings. The LVDT output voltage is a function of the core position:

$$V_m = E \frac{[M'(x) - M''(x)]}{R_1 + L_1 j\omega} j\omega,$$

$M'(x)$ and $M''(x)$ are the mutual inductance between the primary winding and each of the secondary windings.

It is very important to mention that the variation in geometrical parameters also frequently produces an electrical property change. It is therefore often necessary to take into account this interdependence, which can fundamentally affect the determination of the transducer sensitivity.

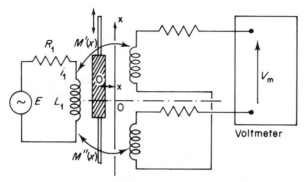

Figure 2.2 A LVDT is an inductive transucer

Example

The strain gauge is a passive transducer that converts a mechanical displacement into a change of resistance. It is a thin, wafer-like device that can be bonded to a variety of materials to measure applied strain (Figure 2.3). The sensitivity of a strain gauge is called the *gauge factor K,* defined as the unit change in resistance (R) per unit change in length:

$$K = \frac{\Delta R/R}{\Delta l/l}.$$

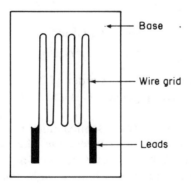

Figure 2.3 Strain gauge is a passive transducer

The resistance change ΔR of a conductor with length l and cross-sectional area S can be calculated by using the following expression:

$$R = \rho \frac{\text{length}}{\text{area}} = \rho \frac{l}{S}.$$

The resistance change can be expressed as follows:

$$\frac{\Delta R}{R} = \frac{\Delta l}{l} - \frac{\Delta S}{S} + \frac{\Delta \rho}{\rho}, \quad (2.1)$$

where $S = (\pi/4)d^2$ (d = diameter of the conductor). Tension on the conductor causes an increase Δl in its length and a simultaneous decrease Δd in its diameter:

$$\frac{\Delta d}{d} = -v\frac{\Delta l}{l},$$

where v = Poisson's ratio. Then we can write:

$$\frac{\Delta S}{S} = -2v\frac{\Delta l}{l} \quad (2.2)$$

and

$$\frac{\Delta \rho}{\rho} = C\frac{\Delta V}{V},$$

where V = volume = Sl and C = Bridgman constant. Therefore:

$$\frac{\Delta \rho}{\rho} = C(1 - 2v)\frac{\Delta l}{l} \quad (2.3)$$

Substitution of Eq. (2.2) and (2.3) into Eq. (2.1) yields:

$$\frac{\Delta R}{R} = [(1 + 2v) + C(1 - 2v)]\frac{\Delta l}{l}. \quad (2.4)$$

The sensitivity of the strain gauge is equal to:

$$K = (1 + 2v) + C(1 - 2v). \quad (2.5)$$

For metallic strain gauges K is of the order of 2 because C is roughly equal to 1. For semiconductor strain gauges, C is about 200, K is between 100 and 200.

Their high sensitivity makes these semiconductor transducers suitable for the measurement of low strain. However as they are characterized by nonlinearity and high drift, care is necessary in their use.

Metallic strain gauges are used where a wide temperature range may occur and for large pressure amplitudes.

2.1.2 Passive Transducer Interfacing

The impedance variation of a passive transducer may be converted into:
— *voltage information* V_m of the measuring signal: in this case the signal conditioner may be a voltage divider, a high-impedance current source or a deflection bridge circuit.

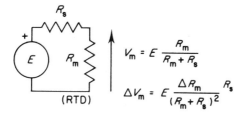

Figure 2.4   Voltage divider

— *frequency information* f_m of the measuring signal: the signal conditioner is an oscillator. This type of information capture is convenient for use in telemetry application.

2.1.2.1   Signal conditioner for voltage information

(1) Voltage-Divider Circuit and Current Source

These circuits are undoubtedly the simplest signal conditioners. The measurand, in effect, modulates the excitation which can be used to provide an increased output level (Figure 2.4 and 2.5).

Consider a simple resistance temperature detector (RTD). For accurate resolution of small temperature changes, with correspondingly small resistance changes, it would be useful to increase the level of the excitation source E or I. However when E and I are sufficiently great, the power dissipated in R_m will itself cause the temperature to rise perceptibly due to self-heating. A measurement error is thus introduced. Furthermore, this type of signal conditioner cannot cope with very small resistance changes. In strain gauges, these changes can be about 10^{-6} ohm that is to say about 1/10 000 ohm. If the strain gauge resistance is equal to 120 Ω this resistance must be measured to with seven significant digits. A possible solution to this problem is to use a deflection bridge circuit.

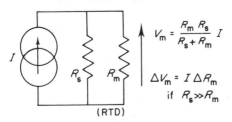

Figure 2.5   Current source

(2) Bridges

In its simplest form, a bridge consists of four passive elements, a source of excitation and a detector as shown in Figure 2.6. It measures the impedance

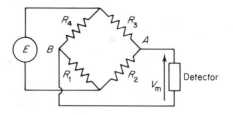

Figure 2.6 Basic bridge circuit

of a passive transducer indirectly by comparison against a similar element. When the bridge is balanced, the potential difference V_m is 0 V; this condition occurs when $R_4 R_2 = R_1 R_3$.

The deviation of one (or more) impedance in the bridge from an initial value must be measured as an indication of a change of the measurand.

Figure 2.7 shows a bridge with all resistances nominally equal, but one of them (R_1) is variable by a factor $R + \Delta R$.

Then the bridge is unbalanced, and we can write:

$$V_A = \frac{E}{2}$$

and

$$V_B = E \frac{R + \Delta R}{2R + \Delta R}$$

The bridge output is equal to:

$$V_m = V_A - V_B = \frac{E}{4} \frac{\Delta R}{R} \frac{1}{1 + \Delta R/2R} \tag{2.6}$$

As the equation indicates, the relationship between V_m and ΔR is not linear, but for small ranges of ΔR it is sufficiently linear for many purposes.

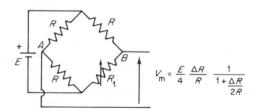

Figure 2.7 Bridge used to read deviation of a passive transducer (R_1)

Example A metallic strain gauge with a gauge factor of $K = 2$ is fastened to a steel member subjected to a strain

$$\sigma = \frac{\Delta l}{l} \leq 5 \times 10^{-3}$$

Therefore:

$$\frac{\Delta l}{l} = \frac{1}{K}\frac{\Delta R}{R} = \frac{\Delta R}{2R} \leq 5 \times 10^{-3} \ll 1$$

The output of the bridge V_m is practically linear with ΔR. If a semiconductor strain gauge is used ($K = 100$) it is necessary to linearize the signal from the bridge circuit, because the non-linearity cannot be neglected since

$$\frac{\Delta R}{2R} = \frac{\Delta l}{2l} K = 2.5 \times 10^{-1}$$

when compared to 1.

Tables 2.3–2.5 show typical bridge configurations used for passive transducers.

Table 2.3   Resistive transducers bridges

Bridge configurations	Comments
Bridge with a single transducer	1. The relationship between V_m and ΔR is not linear. 2. Sensitivity of the bridge: $$\frac{V_m}{\Delta R} \simeq \frac{E}{4R} \text{ for } \frac{\Delta R}{2R} \ll 1$$ 3. The sensitivity can be doubled if two identical transducers can be used, e.g. at position R_3 and R_1.
Single op amp as a bridge amplifier	*Advantage*: simple *Disadvantages*: 1. The external resistances must be carefully chosen and matched to maximize common mode rejection (CMR) which will be discussed in the next chapter. 2. Difficulty to switch the gain.
Instrumentation amplifier applied to bridge measurement	*Advantages*: —low drift —high CMR —high input impedance —when the bridge is at a high potential with respect to the signal conditioning circuitry, an *isolation amplifier* may be used.

Table 2.4 Capacitive transducers bridges

Bridge configurations	Comments
Variation of the distance between two parallel plates $$\Delta C = \frac{\varepsilon_0 S}{d^2} \Delta d$$ 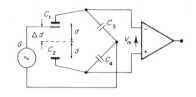 $$V_m = E \frac{\Delta d}{2d}$$ Bridge with two variable elements	1. When $\Delta d = 0$, $V_m = 0$ because the bridge is balanced ($C_1 C_4 = C_2 C_3$) 2. The relationship between V_m and Δd is linear 3. The output is twice that of the single variable element bridges 4. Good resolution
Variation of relative permittivity ε_r 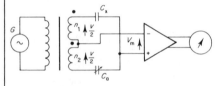 C_x = capacitance transducer n_1, n_2 = winding numbers $$V_m = \frac{V}{2}\left(\frac{1 - C_a/C_x}{1 + C_a/C_x}\right)$$ Single transformer bridge 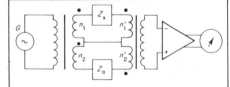 Z_x = transducer at balance: $$\frac{Z_x}{n_1 n_1'} = \frac{Z_a}{n_2 n_2'}$$ Double transformer bridge	—high accuracy (very high-precision transformers are possible today). —Isolation can be provided by using a double transformer bridge. —Typical application: (humidity measurement). Wet material is located between the two capacitance plates. Therefore we can write: $$\frac{1}{Z_x} = \frac{1}{R} + jC\omega$$ R and C depend on humidity material. If only R changes: $$\frac{dZ_x}{Z_x} = \frac{dR}{R} - \frac{jC\omega}{1+jRC\omega}dR = \frac{dR}{R}\left(\frac{1}{1+jQ}\right)$$ with $Q = RC\omega$ If only C changes: $$\frac{dZ_x}{Z_x} = -\frac{dC}{C}\frac{1}{1-j/Q}$$ *Excitation choice*: For high-precision measurements, it is recommended to use a low-frequency source for R measurement, but a high-frequency source for C measurement.

Table 2.5 Inductive transducer bridges

Bridge configurations	Comments
Inductance measurement bridge	Maxwell bridge: $$R_x = \frac{R_2 R_3}{R}; \quad L_x = R_2 R_3 C$$ This bridge is limited to the measurement of medium Q coils ($1 < Q < 10$). The Hay bridge differs from the Maxwell bridge by having R in series with standard C. It is more convenient for measuring high-Q coils.
Differential transformer with an E core	—The armature is rotated about a pivot point. When it is displaced from its balance the reluctance of the magnetic circuit through one secondary coil is decreased while the reluctance of the magnetic circuit through the other secondary coil is increased. —The signal conditioner operates in the same manner as an LVDT.

For the majority of resistive transducers (RTD, thermistor, strain gauge etc.) the excitation supply may be a direct voltage. For high-precision measurement, a stable and accurate supply may be required. Some specified complete signal conditioners provide programmable excitation in addition to amplification.

The hardware to detect and measure the output from a transducer can take many forms. It must be applicable to today's industrial environment which calls for greater speed of response and the ability to interface with analogue or digital signal-handling circuitry. The design task is made much easier by the existence of *signal conditioner packages* which cover the wide range of card options associated with intelligent measurement and control system. These consist of monolithic modules which include an adjustable gain instrumentation amplifier, noise filtering and excitation.

Example 1: Interfacing pressure transducer The output of the transducer must be amplified to a standard 0 to 10 V voltage and then converted to current for transmission in standard 4–20 mA current loops. The strain gauge pressure transducer can be interfaced by a monolithic signal conditioner (Figure 2.8).

Excitation for the bridge, the signal conditioner and the voltage-to-current converter can be provided from a special power supply module. The voltage output of the signal conditioner can be converted to isolated 4–20 mA current for transmission to a recorder.

Analog Devices Inc. manufactures a wide variety of products for these applications. Examples of products in these categories include the model 2B30 signal conditioners, the 2B22 voltage to current-loop converter and the 2B35 transducer power supply.

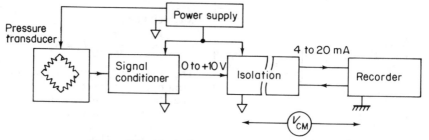

Figure 2.8 Typical pressure measuring system

Example 2: LVDT interfacing In order to obtain useful information from the LVDT, a series of operations is required:
(1) A stable source of constant frequency excitation voltage must be applied to the primary of the LVDT (Figure 2.9).
(2) A full-wave rectifier will provide usable amplitude information; relative phase information is lacking. In order to obtain both phase and amplitude information a synchronous demodulator is needed.
(3) The demodulated signal must be filtered by a low-pass active filter.

The NE5520 is a complete LVDT signal conditioner produced by Signetics. This monolithic driver–demodulator is designed to interface with most LVDTs presently being used in industry.

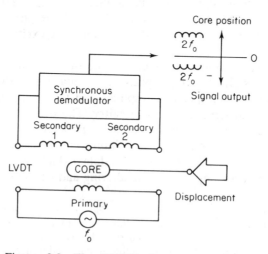

Figure 2.9 The LVDT sine-wave excitation and synchronous demodulation

Perhaps the most widely used form of amplifier for reading bridge outputs is the *instrumentation amplifier* which is characterized by low drift, high common-mode rejection, high input impedance and the capability of maintaining specified performance over a range of gains typically from 1 to 1000.

An isolation amplifier is useful for applications where the bridge may be at a high potential with respect to the signal-conditioning device. Information about instrumentation and isolation amplifiers, in the more general context of measurement systems, will be found in Chapter 4.

2.1.2.2 Signal conditioner for frequency information

This class of signal conditioner converts the impedance variation of a passive transducer into a variation in the frequency of an oscillator circuit.

There are various kinds of oscillator with circuit designs that depend on the frequency they are required to produce.

Low-frequency oscillators operating roughly in the 1 Hz to 20 kHz range are often based on relaxation oscillator circuits.

For higher frequencies, *LC* oscillators can be used. Tables 2.6 and 2.7

Table 2.6 Relaxation oscillators

Configurations	Comments

Single op amp oscillator $$T_m = \frac{1}{f_m} = 2RC \ln\left(1 + \frac{2R_1}{R_2}\right)$$	—Low frequency
	—Low cost
	—*Applications*:
	  Resistive transducer (R)
	  Capacitive transducer (C)
	The op amp can be conveniently replaced by a voltage comparator designed specially to operate from a single power supply over a wide range of voltages (example: LM139 series from National Semiconductor) that allows output voltage to be made compatible with TTL, DTL, ECL, MOS and CMOS logic systems.

Multivibrator using a 555 timer	The 555 monolithic timing circuit is a highly stable oscillator capable of producing accurate time delays or frequency. —High output current (200 mA) —TTL compatible (square wave) —Temperature stability —Frequency: $$f = \frac{1.44}{(R_A + 2R_B)C}$$ *Application*: capacitive transducer (C)

Table 2.7 LC Oscillators

Configurations	Comments
Harley oscillator	—The output signal is available through the transistor collector. Part of the amplified a.c. voltage is fed back to the LC circuit. The feedback causes an output voltage of constant amplitude at the resonant frequency. *Applications*: Capacitive transducer (displacement level, pressure, thickness...)
Colpitts oscillator $f_m = \dfrac{1}{2\pi\sqrt{(LC)}}$ with $\dfrac{1}{C} = \dfrac{1}{C_1} + \dfrac{1}{C_2}$	—The amount of feedback in this oscillator depends on the relative values of C_1 and C_2. *Applications*: Inductive transducer (force, pressure, position, vibration, displacement, temperature,...)

show some of the more usual *LC* and relaxation oscillators used in transducer signal conditioning.

Figure 2.10 shows a typical application of the *LC* oscillator whose frequency is affected by a change in the inductance of the coil. The stability of the oscillator must be excellent in order to detect changes in oscillator frequency caused by the externally applied force. It can be used to measure both static and dynamic phenomena and is convenient for use in telemetry applications.

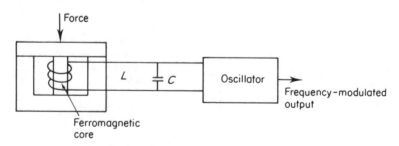

Figure 2.10 Typical application of the *LC* oscillator

Table 2.8

Physical effects	Output signal	Comments
Piezoelectric effect	The quartz cristal produces an *electrical charge* (Q) when weights are placed on it.	*Applications*: —force transducer —pressure —accelerometer —vibration *Features*: —very good high-frequency response —its output signal is easily affected by temperature variations.
Photoelectric effects	The photoemissive cell or photodiode is a radiant energy device that controls its electron emission when exposed to incident light. Consequently, the output signal is an *electrical current* (I_p).	*Application*: —light and radiation
Photovoltaic effect	A *voltage V* is generated in a semiconductor junction device when radiant energy stimulates the cell.	*Applications*: —light meter —solar cell
Electromagnetic effect	Motion of a coil in magnetic field (B) generates a *voltage* (V)	*Applications*: —velocity —vibration
Hall effect	A *potential difference* is generated across a semiconductor plate when magnetic flux interacts with an applied current.	*Applications*: —magnetic flux detector —current transducer —position transducer —velocity transducer
Thermoelectric effect	An *emf* is generated across the junction of two dissimilar metals when that junction is heated.	*Applications*: —temperature —heat flow —radiation *Features*: —small size and wide temperature range —requires reference to a known temperature —nonlinear response —low source impedance (typically 10 Ω)

2.2 ACTIVE OR SELF-GENERATING TRANSDUCERS

2.2.1 Physical Principles

The electrical signal at the output of an active or self-generating transducer is only derived from the physical input; examples include thermocouples, photodiodes, piezoelectric transducer etc. Since the electrical output is limited by the physical measurand, such transducers tend to have low-energy outputs requiring amplification.

Table 2.8 shows the most commonly used physical effects. A large number of useful phenomena and their sensors will be omitted as they do not have such widespread application and also because this book is principally intended for use with electronic circuits for transducer interfacing. In order to design an adequate signal conditioner, it is essential to know the nature of the transducer output signal.

2.2.2 Active Transducer Interfacing

These active transducers can be modelled as:

— a *Voltage source* (thermoelectric effect, Hall effect, electromagnetic effect, photovoltaic effect, . . .)
— a *current source* (photoelectric effect, piezoelectric effect, . . .)

which require special networks for signal conditioning.

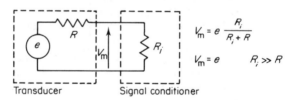

Figure 2.11 Equivalent circuit of a voltage source transducer

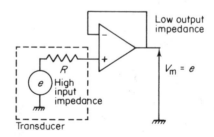

Figure 2.12 Unity-gain follower

2.2.2.1 Voltage-source transducers

Figure 2.11 shows the equivalent circuit of a voltage-source transducer. It usually requires amplifiers for buffering, isolation and gain. Most of these functions can be performed by operational amplifiers (op amp).

The op amp is an excellent unity gain follower (Figure 2.12). If the feedback is attenuated we have a follower with gain where the gain is accurately set by a resistances ratio (Figure 2.13).

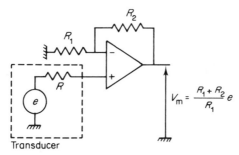

Figure 2.13 Follower with gain

2.2.2.2 Current-source transducers

Figure 2.14 shows the equivalent circuit of a current-source transducer which provides a current output i_m through the signal conditioner input impedance R_i. Usually these transducers require a current-to-voltage conversion, which can be performed by the very simple op amp circuit given in Figure 2.15. The output voltage is independent of R.

It is interesting to note that the output of a piezoelectric transducer may be modelled as a current source in parallel with a small capacitor (Figure 2.16).

The forces to be measured are dynamic (i.e. continually changing over the period of interest). Charge amplifier configurations are required for signal conditioning. In general, manufacturers of piezoelectric devices furnish calibrated charge amplifiers and cables. The functional description of charge amplifiers will be discussed in Chapter 4.

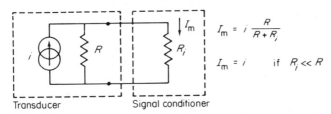

Figure 2.14 Equivalent circuit of a current-source transducer

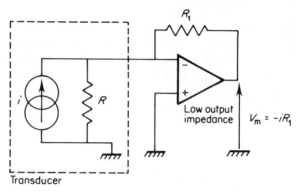

Figure 2.15 Current-to-voltage converter

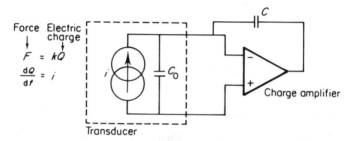

Figure 2.16 Piezoelectric transducer modelled as a current source

2.3 OP AMP SELECTION PROCESS

The op amp has become one of the most versatile elements in transducer interfacing. Most designers understand its basic properties and applications, but the subtleties of using op amps for best results especially in precision measurement and control are not sufficiently understood by many op amp users.

To assist the designer in distinguishing among the many types of op amp we can classify them as shown in Table 2.9.

There can be overlap between some categories: some types can cover both high accuracy and low power. However, it is difficult to simultaneously optimize both high speed and low power.

Economics are commonly an important dimension of the choice process. A high-accuracy op amp may be more cost effective if it eliminates the need for external trimming components. Some general-purpose types are often used in place of high-accuracy devices because of their low input bias currents and low input offset voltage. The need for low bias current can be met through the use of general-purpose FET-input op amps (LF 356 National Semiconductor), but the current rises with increasing temperature.

Table 2.9

Category	Typical characteristics	Applications
General purpose	Lowest cost op amp Low slew rate (<10 V/μs) Moderate offset voltage Moderate bias current Model: 741	General purpose designs Impedance buffering Active filtering Unity gain follower Current-to-voltage converter
High accuracy	Low input offsets (<1 mV) Low drift (<2 μV/°C) Low input bias (<200 pA) High d.c. gain High common mode rejection (>100 dB) Low input noise (<15 nV/$\sqrt{Hz}$) Models: 52 K (Analog Devices) SE/NE 5512 (Signetics) OP-07 (Precision Monolithics Incorporated)	Transducer interfacing designs Amplifier for bridge measurement Charge amplifier for piezoelectric accelerometers Precision integration. Amplification of microvolt level signals. Measurement with very high internal impedance transducers.
Low power Wide supply range	Low supply drain (<1 mA) Wide input and output voltage range High open loop gain Low input offsets High CMRR Models: OP 20/21/22 (PMI, Precision Monolithics Incorporated)	Battery-powered applications Single supply voltage applications Portable instruments Missiles, spacecraft
High speed	High slew rate (<100 V/μs) High-gain bandwidth (>10 MHz) Fast settling time (<500 ns) Models: 50 J/K (Analog Devices) ICL 8017 (Intersil)	High-frequency active filter High-frequency oscillators Fast integrators This category is not commonly required for transducer interfacing

2.4 LINEARIZING

A linear transducer is one for which output signal is proportional to measurand. Figure 2.17 shows an input–output plot of an ideal linear relationship, a nonlinear relationship and the difference between the two (nonlinearity).

In this section we shall discuss some usual linearizing processes that are often found necessary in making sensitive and accurate measurements.

To linearize the signal from a transducer, two principal techniques can be

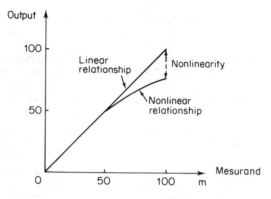

Figure 2.17 Typical nonlinear relationship

employed:

(1) modification of the transducer circuitry;
(2) processing of the transducer's output signal.

2.4.1 Modification of the Transducer Circuitry

An example of modifying transducer circuitry is the use of networks involving thermistors and resistors to obtain an output that is linear over limited ranges.

Thermistor manufacturers use this technique to provide linearized devices having linearities to within 0.2°C over ranges such as 0–120°C (Figure 2.18).

Another example of this technique is illustrated in Figure 2.19, where an op amp provides a feedback signal that balances the bridge to obtain an output proportional to the impedance change of the passive transducer (R). At balance we have

$$R = R_1 = R_2 = R_3.$$

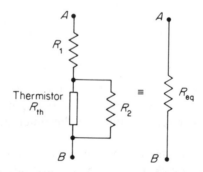

Figure 2.18 Example of thermistor linearizing

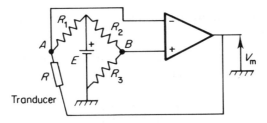

Figure 2.19 Feedback provided by the use of an op amp to linearize the bridge output

When R changes from R to $R + \Delta R$, we can write:

$$V_B = \frac{E}{2}$$

$$V_A = E\frac{R + \Delta R}{2R + \Delta R} + V_m\frac{R}{2R + \Delta R}.$$

Because V_A is always equal to V_B, we obtain:

$$V_m = -\frac{E}{2}\frac{\Delta R}{R} \qquad (2.7)$$

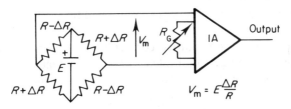

Figure 2.20 Bridge with four variable elements

A final example of modifying the transducer circuitry is shown in Figure 2.20, which shows a deflection bridge consisting of four resistors, two of which increase and two of which decrease in the same ratio. These resistors may be identical strain gauges attached to opposite faces of a beam or flexure to measure its bending. The output voltage of such a bridge would be four times the output for a single transducer bridge.

2.4.2 Processing of the Transducer Output Signal

This technique is used to compensate for nonlinearity in the transducer or the associated signal conditioner. It depends on the nature of the nonlinearity.

2.4.2.1 Nonlinearity correction using a logarithmic or exponential converter

A logarithmic converter provides an output voltage proportional to the logarithm of the input voltage. If V_m is an exponential function of the measurand m,

$$V_m = V_0 e^m \longrightarrow \boxed{\text{Logarithmic converter}} \longrightarrow s = -K \log \frac{V_m}{V_{\text{ref}}}$$

this logarithmic device will provide a linear output; as a matter of fact, the output voltage can be written:

$$s = -K \log \frac{V_m}{V_{\text{ref}}} = -K \log \frac{V_0 e^m}{V_{\text{ref}}}.$$

If $V_{\text{ref}} = V_0$, we obtain

$$s = -Km.$$

Usually the logarithmic device can be connected for exponential operation (Chapter 5). For a logarithmic function input, the exponential converter will give a linear output.

2.4.2.2 Nonlinearity correction using an analogue multiplier

An analogue multiplier is a device which provides an output voltage equal to the product of two input voltages, multiplied by a scale constant (Chapter 5).

An example of a nonlinearity correction for a large-deviation off-null bridge using an analogue multiplier is shown in Figure 2.21. The output voltage of single transducer bridge (Eq. 2.5) is equal to:

$$V_m = \frac{E}{4} \frac{\Delta R}{R} \frac{1}{1 + \Delta R/2R}, \qquad (2.8)$$

which is not directly proportional to ΔR.

In Figure 2.21, the output of the summing device can be written

$$V_0 = b \frac{V_m V_s}{K} + a V_m. \qquad (2.9)$$

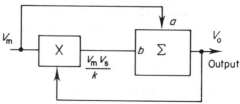

Figure 2.21 Nonlinearity correction using an analogue multiplier

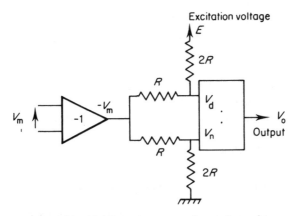

Figure 2.22 Nonlinearity correction using an analogue divider

Substituting Eq. (2.8) into Eq. (2.9), we obtain

$$V_0 = a \frac{E}{4} \frac{\Delta R}{R} \frac{1}{1 + (\Delta R/2R)(1 - bE/4K)}. \qquad (2.10)$$

If we adjust b (the gain of the summing device) in order to have $b = 4K/E$, we can obtain:

$$V_0 = a \frac{E}{4} \frac{\Delta R}{R}. \qquad (2.11)$$

Notice that measurement accuracy depends on the power supply stability; so it is possible to abolish this deficiency by using a divider as shown in Figure 2.22.

From Figure 2.22, we can write:

$$V_n = -\tfrac{2}{3} V_m$$

$$V_d = -\tfrac{2}{3} V_m + \frac{E}{3}$$

and

$$V_0 = K \frac{V_n}{V_d} = \frac{-2V_m}{-2V_m + E} K;$$

therefore

$$V_0 = -K \frac{\Delta R}{2R}. \qquad (2.12)$$

Hence the output V_0 is independent of E.

2.5 MONOLITHIC TRANSDUCERS

As technology expands the increased demand for more elaborate and more accurate transducers has led to the development of new transducer designs and applications. Many manufacturers are developing and introducing monolithic transducers that greatly reduce unit cost and allow the electronic designer greater freedom in implementing transducer circuits. In this section we would like to mention some typical monolithic devices in order to indicate how this new technology can simplify the subsequent signal conditioning and processing.

2.5.1 Integrated Circuit Temperature Transducers

These use the fundamental property of silicon transistors to exhibit a temperature proportional characteristic. If two identical transistors are operated at a constant ratio of collector current densities, then the difference in their base–emitter voltages is proportional to KT/q. Since both K, Boltzman's constant and q, the charge of an electron, are constant, the resulting voltage is directly proportional to absolute temperature. Such a transducer is shown in Figure 2.23, where we can write

$$i_1 = i_2 \quad \text{because} \quad i_{C_3} = i_{C_4},$$

$$V_{BE}(T_2) = R i_1 + V_{BE}(T_1) = \frac{KT}{q} \ln \frac{i_2}{i_{S_2}}$$

$$V_{BE}(T_1) = \frac{KT}{q} \ln \frac{i_1}{i_{S_1}}.$$

therefore:

$$R i_1 = \frac{KT}{q} \ln \left(\frac{i_2}{i_{S_2}} \cdot \frac{i_{S_1}}{i_1} \right) = \frac{KT}{q} \ln r \frac{i_2}{i_1} \quad \text{with} \quad \frac{i_{S_1}}{i_{S_2}} = r.$$

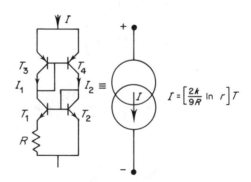

Figure 2.23 Simplified functional diagram of integrated circuit temperature transducer

Finally we have:

$$i_1 = \frac{KT}{Rq} \ln r;$$

then

$$i = i_1 + i_2 = 2i_1 = \left(\frac{2K}{qR} \ln r\right) T$$

Numerical example

$r = 8, \quad K = 1.38 \times 10^{-23} \text{ J/K}, \quad q = 1.6 \times 10^{-19} \text{ C}.$

If the transducer sensitivity is equal to $1 \mu A/K$, *we must have*

$R = 359 \, \Omega.$

The transducer is modelled by a simple current source.

Example of industrial product
Two-terminal IC temperature transducer
AD 590 M (Analog Devices)

Features:

Current output: $1 \mu A/K$
wide range: $-55°C$ to $+155°C$
excellent linearity: $+0.3°C$ over full range

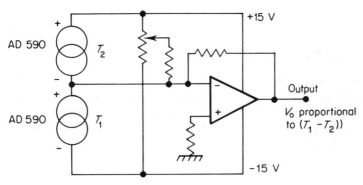

Figure 2.24 ΔT measurement using AD 590 Transducers

For applications calling for voltage readout, the circuit of Figure 2.24 may be used. It indicates one way in which differential measurements can be made using two AD 590 devices and a single op amp.

2.5.2 Monolithic Pressure Transducers

In general, monolithic pressure transducers are piezoresistive integrated circuits which provide an output voltage proportional to applied pressure. The devices are provided in compact package with pressure ports.

Advantages of Monolithic Transducers

The monolithic transducer is easily temperature compensated with respect to sensitivity and features low offset temperature coefficient. High sensitivity and low noise allow easy amplification.

The most effective bridge buffering circuits are instrumentation amplifiers which provide very high input impedance. It is possible to convert the output voltage into frequency output compatible with an analogue-to-digital converter or microprocessor input.

Example of industrial products Monolithic pressure transducers (National Semiconductor) (Table 2.10)

Table 2.10

Series	Type	Sensitivity	Operating pressure range
LX 0503 A	Absolute pressure transducer	2–8 mV/psi	0 to 30 psid
LX 0603 D	Differential pressure transducer	2–8 mV/psi	−30 psid to +30 psid

2.5.3 Hall-effect Integrated Circuits

These devices include a monolithic Hall cell, linear differential amplifier, differential emitter follower output and a voltage regulator. Integrating the Hall cell and the amplifier into one monolithic device minimizes problems related to the handling of millivolt analogue signals. The differential output of the devices is a function of the magnetic flux density present at the

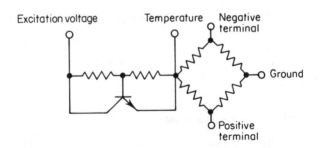

Figure 2.25 Schematic diagram of a monolithic pressure transducer

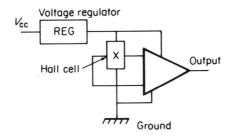

Figure 2.26 Functional diagram of Hall effect integrated circuit

sensor. Sensitivity, which is a function of the control current, increases as the control current increases.

The Hall effect initially found limited application in wattmeters or gaussmeters, because of its complexity, cost and susceptibility to noise and temperature variations. Production of Hall-effect integrated circuits have eliminated these problems. These devices are intended for applications requiring accurate measurement of position, weight, velocity, etc. They provide a linear output voltage which is a function of magnetic field intensity, which may be produced by the displacement of a permanent magnet.

Example of industrial products
  Type: UGN-3501 M (SPRAGUE)
  Typical sensitivity: 1.4 V/1000 gauss
  frequency response: 25 kHz.

Chapter 3
PRACTICAL ASPECTS OF NOISE REDUCTION IN ELECTRONIC MEASURING SYSTEMS

3.1 DEFINITIONS

Electromagnetic interference is presently a very difficult problem for measuring-systems designers and is likely to become more difficult in the future because a larger number of electric systems for communication, power distribution, automation, etc. have to operate in close proximity and so affect each other adversely. Elimination of electromagnetic interference has become a major objective in the design of measurement systems.

Noise can be defined as any electrical signal present in a circuit other than the desired signal. It can be produced by fundamental fluctuations within electronic systems due to thermal noise, shot noise, active device noise, etc or by electromagnetic radiation from various external sources such as motors, switches, transmitters, power distribution systems, etc.

Interference is considered as the undesirable effect of noise causing unsatisfactory measurements of the useful signal. Practically it is not possible to eliminate noise, but it is possible to reduce its magnitude until it no longer causes interference,

Three steps are necessary in analysing a noise problem:

(1) determination of the noise source;
(2) determination of the coupling channel to transmit the noise from its source to the measuring equipment;
(3) determination of the measuring circuit that is susceptible to the noise.

It follows that if the noise cannot be reduced at the source, then the coupling channel must be suppressed and/or the measuring system must be made insensitive to the noise.

It should be remembered that a universal solution to the noise-reduction problem does not exist. Compromises, which are specific to the application are generally required. In this book a simple presentation of the more usual techniques, which are useful for decreasing interference in measuring systems will only be given. Obviously, the main methods by which noise coupling can be eliminated, such as shielding, grounding and balancing, will be considered. Other techniques, which depend upon filtering, isolation, power-supply decoupling, etc, will be mentioned at appropriate points throughout the book.

3.2 EXAMPLES OF NOISE COUPLING

3.2.1 Capacitive Coupling

A simple example of capacitive coupling is shown in Figure 3.1. Conductor 1 is considered as the noise source, and conductor 2 as the affected circuit. V_1 is the voltage on conductor 1. C_{12}, C_{1G}, and C_{2G} are stray capacitance. The noise voltage V_2 produced on conductor 2 is equal to:

$$V_2 = \frac{C_{12}}{C_{12} + C_{2G}} \frac{1}{1 + \dfrac{1}{Rp(C_{12} + C_{2G})}} V_1. \quad (3.1)$$

V_2 is maximum if $\omega > \omega_0$ (Figure 3.2). In most practical cases, this equation can be reduced to the following:

because
$$V_2 = pRC_{12}V_1 \quad \text{with} \quad p = j\omega \quad (3.2)$$

$$R \ll \frac{1}{j\omega(C_{12} + C_{2G})}.$$

Numerical example
$R = 50\,\Omega$, $C_{12} = 50\,\text{pF}$, $C_{2G} = 150\,\text{pF}$
$V_1 = 10\,\text{V}$, $f = 100\,\text{kHz}$

$$\frac{1}{j\omega(C_{12} + C_{2G})} = \frac{1}{2\pi \times 10^5 (50 + 150)10^{-12}} = 8 \times 10^3\,\Omega \gg R.$$

Magnitude of noise voltage:
$V_2 = 2\pi \times 10^5 \times 50 \times 50 \times 10^{-12} \times 10$
$V_2 = 15.7\,\text{mV}.$

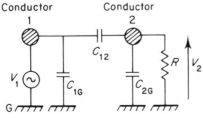

Figure 3.1 Capacitive coupling between two conductors

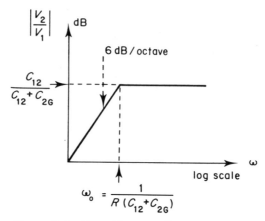

Figure 3.2 Capacitive coupling is maximum in high frequency

Equation (3.1) shows that the magnitude of noise voltage is proportional to the frequency of V_1, the stray capacitance C_{12}, the resistance R and the magnitude of V_1. Assuming that the noise source and the resistance R cannot be changed, this leaves only one remaining parameter for reducing capacitive coupling: capacitance C_{12} can be reduced by separating the conductors or by shielding.

Consider the case where a shield is placed around conductor 2, but in practice the centre conductor extends beyond the shield and the configuration becomes that of Figure 3.3. If the shield is properly grounded the noise voltage V_2 is the same as Eq. (3.1), which is for an unshielded cable, except that C_{12} is greatly reduced, because it consists of capacitance between conductor 1 and the unshielded portion of conductor 2.

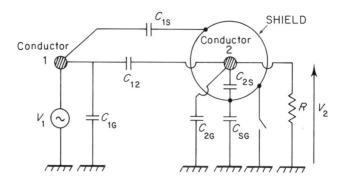

Figure 3.3 Effect of shield on capacitive coupling

Numerical example
$R = 50\ \Omega$, $V_1 = 10$ V, $f = 100$ kHz
$C_{12} = 2$ pF
Magnitude of noise voltage:
$V_2 = 2\pi f R C_{12} V_1$
$V_2 = 630\ \mu$V.

3.2.2 Inductive Coupling

Consider a magnetic field of flux density B sinusoidally variant with time. The voltage V_2 induced in a closed loop of area A due to B is equal to:

$$V_2 = pBA \cos \theta.$$

If the magnetic field is produced by a current I_1 in the conductor 1 (Figure 3.4), Eq. (3.3) can be expressed as follows:

$$V_2 = pMI_1.$$

V_2 is the noise voltage produced in the conductor 2 by the noise source $\dot{I}_1$ (which is proportional to V_1).

To reduce the noise voltage:

(1) Conductor 2 must be separated from the noise source conductor.
(2) The magnetic field of density B cuts area A at an angle of θ; the $\cos \theta$ term can be reduced by proper orientation of the receiver conductor.
(3) The area of the receiver circuit must be minimized by placing the conductor 2 closer to the ground plane. This technique is efficient only when this ground plane is crossed by the return current.
(4) The use of a twisted conductor pair for the receiver conductor may be very efficient if the return current flows in one of the pair instead of the ground (Figure 3.5). The maximum frequency in this case is about

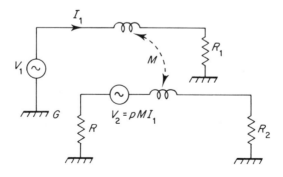

Figure 3.4 Inductive Coupling

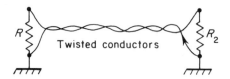

Figure 3.5 The return current is on one of the pair instead of the ground

100 kHz. For higher frequency (up to 1 GHz) coaxial cable must be used.

However, electronic system design requires further care. Grounding a circuit at both ends should be avoided because of the noise current induced in the ground loop.

An example of the effect of noise current flowing in the shield of a coaxial cable is shown in Figure 3.6.

The current noise I_s may be produced by a ground differential (between G_s and G_R) or by magnetic coupling. An undesired voltage is then applied at the amplifier or voltmeter input:

$$V_{in} = R_s I_s$$

because in the entire expression:

$$V_{in} = -j\omega M I_s + j\omega L_s I_s + R_s I_s. \tag{3.4}$$

The mutual inductance between the shield and the centre conductor is equal to the shield inductance $(M = L_s)$. Therefore, for maximum noise protection

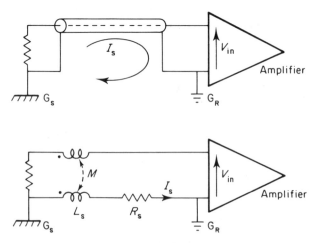

Figure 3.6 Effect of noise current in a circuit grounded at both ends

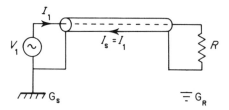

Figure 3.7 One end of the system is isolated from ground

one end of the system must be isolated from ground (Figure 3.7). However this technique is valid only at low frequencies. At high frequencies it must be remembered that noise voltages may be produced by capacitive coupling through the coaxial cable.

3.2.3 Conductive Coupling

An ideal conductor does not exist. Whenever a current flows through a conductor, it produces a voltage drop and causes interference.

We have seen in Figure 3.6 that in a circuit grounded at both ends there may be a ground loop which is susceptible to differences in ground potential. It also causes interference.

As grounding is one of the primary ways to reduce interference, the main objectives for a good grounding circuit design are important.

Generally, a ground must be defined as an equipotential plane which serves as a reference voltage for signal voltage. This is called the *signal ground* and it may be at earth potential. If the signal ground is connected to the earth through a conductor it is then called an *earth ground*.

3.3 GROUNDING CONNECTIONS

Three usual grounding connections are shown in Figure 3.8.

In the following discussion it should be remembered that all conductors have a finite impedance which generally consists of both resistance and inductance.

Connection (a) This is obviously the most widely used because of its simplicity. As it is the most noisy ground system, this connection must be avoided. The low end of circuit (3) is at about 30 mV and the low end of low power circuit (1) is at about 10 mV. If, for example, (1) is a high-accuracy op amp with its positive input connected to ground, the summing point will effectively be at an offset of 10 mV with respect to a signal source referenced to the power supply.

Connection (b) One approach to improve the situation is shown in (b); a separate lead is run from each circuit to the low end of the power supply.

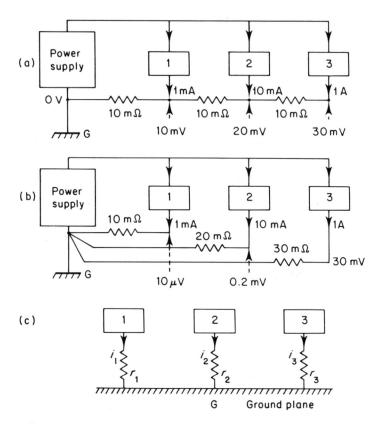

Figure 3.8 Different grounding connections

The offset at circuit (1) is now negligible. But the offset at (3) is still about 30 mV; Further improvement for (3) may be the use of a separate supply.

Connection (c)   This is specially used at high frequencies (>10 MHz). The connections should be very short. If required the common impedance of the ground plane can be reduced by silver plating the surface.

3.4 GROUNDING OF CABLE SHIELDS

In this grounding technique, it is important to emphasize that two separated points are seldom at the same potential. If the shield is grounded at two points (Figure 3.6) noise current will flow and cause a voltage drop in the shield resistance. To avoid this noise current, it is necessary to ground the shield at only one point. Consider an amplifier shown in Figure 3.9: this system consists of an ungrounded source and a grounded amplifier.

Connection A: This allows noise current to flow in one of the signal leads to return to ground, and produces a noise voltage in series with the useful signal V_s.

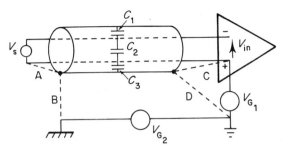

Figure 3.9 Different connections in an amplifier system

Connection B: C_1 and C_2 form a capacitive voltage divider. Assuming $V_s = 0$ then a voltage is generated across the amplifier input:

$$V_{in} = \frac{C_1}{C_1 + C_2}(V_{G1} + V_{G2}). \quad (3.5)$$

Connection C: $V_{in} = 0$. This connection is satisfactory.

Connection D: $V_{in} = \dfrac{C_1}{C_1 + C_2} V_{G1}.$ (3.6)

Thus for good grounding of cable shield, it is necessary to connect the input shield to the amplifier common terminal.

In the case of a differential amplifier, when the source is grounded, the best shield connection is shown in Figure 3.10, where the shield is connected to the source common.

When a signal circuit is grounded at both ends, the ground loop formed is susceptible to noise from magnetic fields and differential ground voltages. Techniques for breaking the ground loop are provided by isolation transformers, optical couplers and guarded amplifiers, which will be discussed in Chapter 4.

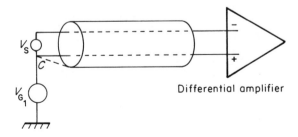

Figure 3.10 Grounding connection in the case of a grounded source and a differential amplifier

3.5 BALANCING TECHNIQUE

3.5.1 Principle of Operation

Generally this technique may be used when noise must be reduced to a very low level. The ideal differential amplifier previously shown in Figure 3.10 is basically a balanced system.

The basic balanced configuration shown in Figure 3.11 consists of a balances source (V_{s1} and V_{s2}) and a balanced load (Z_{L1} and Z_{L2}).

Capacitive coupling effect The noise voltages produced in conductors 1 and 2 due to the generator V_3 are equal to

$$V_A = V_3 \frac{Z_{L1}}{Z_{L1} + 1/C_{31}p} \tag{3.7}$$

and

$$V_B = V_3 \frac{Z_{L2}}{Z_{L2} + 1/C_{32}p} \tag{3.8}$$

since $C_{31} = C_{32}$ and $Z_{L1} = Z_{L2}$ because the circuit is balanced, then:

$$V_A - V_B = 0.$$

Inductive coupling effect Generators V_1 and V_2 represent inductive coupling. In this balanced circuit, I_1 is equal to I_2, so

$$V_A - V_B = 0.$$

Conductive coupling effect The difference in ground potential between source and load is represented by V_G. It has no influence on V_A and V_B.

Only useful signals V_{s1} and V_{s2} can generate the desired difference voltage on the differential amplifier inputs, if the measuring system is perfectly balanced.

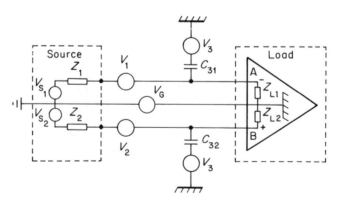

Figure 3.11 Different noise voltages in a balanced configuration

3.5.2 Common-mode Noise Voltage

The use of a differential amplifier, as previously indicated, is the first step toward a balanced measuring circuit. What is the degree of balance of a differential amplifier? What is the degree of balance of a measuring circuit associated with an ideal differential amplifier?

The symmetrical differential amplifier is a basic building block of linear integrated circuits. Its wide acceptance is due to its symmetrical design and excellent performance to direct and low frequency current. The fact that matched transistors with nearly identical parameters are available is one of the many advantages offered by monolithic technology.

Since a differential amplifier has two inputs and two outputs, the two-part representation used for single transistor amplifiers is not applicable. Figure 3.12 shows a differential amplifier controlled by two voltages V_{i1} and V_{i2} which can be decomposed into a *differential input voltage*:

$$V_{id} = V_{i2} - V_{i1} \qquad (3.9)$$

and a *common input voltage*:

$$V_{ic} = \frac{V_{i1} + V_{i2}}{2}. \qquad (3.10)$$

In most practical cases, $V_{02} = 0$, the output voltage will be equal to $V_{01} = V_0$.

The transfer function of an ideal differential amplifier can be written as follows:

$$V_0 = G(V_{i2} - V_{i1}). \qquad (3.11)$$

Because of the inevitable asymmetry of the device, each input component is differently amplified, so the output can be related to the input voltage components as:

$$V_0 = A_2 V_{i2} - A_1 V_{i1}.$$

Employing Eqs. (3.9) and (3.10), we have:

$$V_{i2} = V_{ic} + \frac{V_{id}}{2}$$

$$V_{i1} = V_{ic} - \frac{V_{id}}{2}.$$

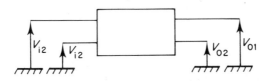

Figure 3.12 Differential Amplifier

Expressing the output voltage in terms of V_{ic} and V_{id} we obtain

$$V_0 = A_2\left(\frac{V_{id}}{2} + V_{ic}\right) - A_1\left(-\frac{V_{id}}{2} + V_{ic}\right)$$

$$V_0 = \frac{A_2 + A_1}{2} V_{id} + (A_2 - A_1)V_{ic}$$

$$V_0 = \frac{A_2 + A_1}{2}\left(V_{id} + 2\frac{A_2 - A_1}{A_2 + A_1} V_{ic}\right). \tag{3.12}$$

The differential voltage amplification can be defined as follows:

$$G_d = \frac{A_1 + A_2}{2}. \tag{3.13}$$

Since V_{id} is the desired input voltage, Eq. (3.12) shows an undesired term:

$$2\left(\frac{A_2 - A_1}{A_2 + A_1}\right)V_{ic}, \tag{3.14}$$

which characterizes the *common-mode rejection ratio* (CMRR):

$$\text{CMRR} = \frac{A_1 + A_2}{2(A_2 - A_1)}. \tag{3.15}$$

With this definition, the output voltage can be expressed by

$$V_0 = G_d\left(V_{id} + \frac{1}{\text{CMRR}} V_{ic}\right) \tag{3.16}$$

As evident from Eq. (3.16) the rejection will be effective only if the common voltage amplification $G_c = A_2 - A_1$ of the amplifier is much lower than its differential voltage amplification $G_d = (A_1 + A_2)/2$. An ideal differential amplifier is characterized by

$$A_2 = A_1 \to \text{CMRR} = \infty.$$

An actual differential amplifier may be represented by the equivalent scheme shown in Figure 3.13.

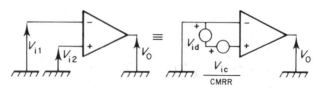

Figure 3.13 CMRR Effect of a Differential Amplifier

3.5.2.1 Effect of input and source impedance asymmetry

Assume that the amplifier CMRR is infinite and the input impedances have finite values (Figure 3.14). The differential input voltage can be written as

$$V_{id} = -V_{i1}\frac{Z_{CH}//(Z_d + Z_{CB}//R_{S2})}{R_{S1} + [Z_{CH}//(Z_d + Z_{CB}//R_{S2})]}$$
$$+ V_{i2}\frac{Z_{CB}//(Z_d + Z_{CH}//R_{S1})}{R_{S2} + [Z_{CB}//(Z_d + Z_{CH}//R_{S1})]}.$$

Assume that $Z_{CB} = Z_{CH} \gg Z_d$ and $Z_{CB} = Z_{CH} \gg R_{S1}$ and R_{S2} then:

$$V_{id} = -V_{i1}\frac{Z_d + R_{S2}}{R_{S1} + Z_d} + V_{i2}\frac{Z_d + R_{S1}}{R_{S2} + Z_d}.$$

If the differential input voltage, which is also the useful input voltage, is zero, that is to say

$$V_{i2} - V_{i1} = 0 \rightarrow V_{i1} = V_{i2} = V_{ic},$$

then the output voltage should be zero, but unfortunately there is always an inevitable error voltage:

$$V_{id} = \frac{(Z_d + R_{S1})^2 - (Z_d + R_{S2})^2}{(Z_d + R_{S2})(Z_d + R_{S1})}V_{ic}$$

$$V_{id} = \frac{(2Z_d + R_{S1} + R_{S2})(R_{S1} - R_{S2})}{(Z_d + R_{S2})(Z_d + R_{S1})}V_{ic}. \qquad (3.17)$$

Assume that $Z_d \gg R_{S1}$ and $Z_d \gg R_{S2}$, Eq. (3.17) will be reduced as

$$V_{id} = \frac{2(R_{S1} - R_{S2})}{Z_d}V_{ic}.$$

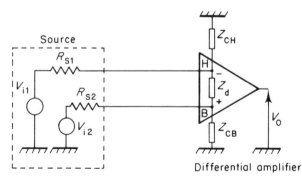

Figure 3.14 Effect of input and sources impedances asymmetry

It is possible to define a common voltage amplification as

$$G_c = \frac{V_0}{V_{ic}} = \frac{2G_d(R_{S1} - R_{S2})}{Z_d}.$$

Then

$$\text{CMRR} = \frac{G_d}{G_c} = \frac{Z_d}{2(R_{S1} - R_{S2})}. \tag{3.18}$$

This ratio represents the degree of balance of the measuring circuit.

Differential amplifier manufacturers specify, in their data sheets, the amplifier CMRR for an unbalanced source impedance equal to $R_{S1} - R_{S2} = \Delta R_S$. Generally the CMRR is expressed in dB (decibels):

$$\text{CMRR}_{dB} = 20 \log_{10} \frac{Z_d}{2(R_{S1} - R_{S2})}. \quad (dB)$$

Numerical example
$Z_d = 100 \text{ M}\Omega$, $\Delta R_S = 1000 \text{ }\Omega$

$$\text{CMRR}_{dB} = 20 \log_{10} \frac{100 \times 10^6}{2 \times 10^3} = 94 \text{ dB}.$$

It is important to bear in mind that we have assumed that the common-mode rejection ratio of the amplifier is infinite. The CMRR previously calculated is only due to unbalanced source impedances. The better the balance, the greater the common noise voltage reduction obtainable.

3.5.2.2 Influence of shielded cables

Usually signal sources are connected to the amplifier inputs by shielded cables which can have different lengths. Figure 3.15 shows stray capacitances between the amplifier input terminals and ground including the cable capacitance.

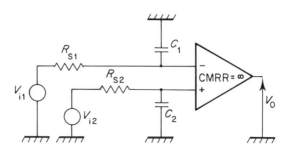

Figure 3.15 Differential amplifier and grounded sources are connected by shielded cables

The difference of capacitances between C_1 and C_2 reduces the degree of balance of the measuring circuit. Assume that $R_{S1} = R_{S2} = R_S$. Then the differential input voltage is

$$V_{id} = V_{i2} \frac{1}{R_S C_2 p} - V_{i1} \frac{1}{1 + R_S C_1 p}$$

if

$$V_{i1} = V_{i2} = V_{iC}$$

$$V_{id} = \frac{R_S(C_1 - C_2)p}{(1 + R_S C_1 p)(1 + R_S C_2 p)} V_{ic}$$

then

$$G_c = \frac{G_d V_{id}}{V_{ic}}$$

and

$$\text{CMRR} = \frac{V_{ic}}{V_{id}} = \frac{(1 + R_S C_1 p)(1 + R_S C_2 p)}{R_s(C_1 - C_2)p} \tag{3.19}$$

Numerical example
$R_S = 10\,\text{k}\Omega$, $C_1 = 110\,\text{pF}$, $C_2 = 100\,\text{pF}$
at $f = 150\,\text{Hz}$, we have:
$R_S C_1 \omega$ and $R_S C_2 \omega \ll 1$

$$\text{CMRR} = \frac{1}{R_S(10 \times 10^{-12})2\pi 150} \approx \frac{1}{10^{-4}}$$

$\text{CMRR}_{dB} = 80\,\text{dB}$.

The common-mode noise voltage can be decreased by proper shielding and grounding.

When extremely low-level signals are being measured or when very large common mode voltages are present, it is necessary to use guard shield techniques, which will be discussed in the following chapter.

3.6 BYPASSING POWER SUPPLIES

In electronic measuring systems, it is usual that the d.c. power supply is common to many circuits. It is necessary to avoid noise coupling between these circuits. An ideal power supply must be a zero-impedance source of voltage. Unfortunately not only does the power supply have a source impedance, but the conductors used to connect it to other circuits add to this impedance. Figure 3.16 shows a typical power-distribution system where the transmission line used to connect the power supply to the load is represented by components R_T, L_T, C_T.

First, it is possible to design the power system so that d.c. voltage drop

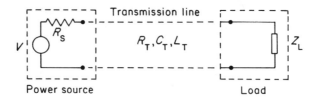

Figure 3.16 Direct current power distribution system

determined by R_S and R_T is minimized: R_S can be reduced by improving the regulation of supply, R_T can be reduced by using very short connection line ($R_T = \rho l/A$). So R_S and R_T determine the d.c. performance of the power distribution system.

Second, it is necessary to reduce noise voltages specially caused by sudden changes in the current demand of the load or by high switching speed in the logic circuits and digital systems due to logic gates drawing a current I_{ON} from the d.c. supply in their ON state and a current I_{OFF} in their OFF state. This current change $\Delta i = I_{ON} - I_{OFF}$ which takes place during Δt seconds generates a noise voltage, across the supply wiring inductance L_T, given by

$$\Delta V_n = L_T \frac{\Delta I}{\Delta t}.$$

This noise voltage may become a major problem if the number of logic gates in the system is large.

Generally, the magnitude of the noise voltage depends on the characteristic impedance (Z_c) of the connection line:

$$Z_c = \sqrt{\left(\frac{L_T}{C_T}\right)}.$$

It is obvious that a good connection line should have a very low characteristic impedance (typically a few ohms); that is to say it should have high capacitance and low inductance.

A low inductance can be obtained by using flat conductors while a high capacitance can be obtained by having the two flat conductors as close together as possible with a high dielectric constant material (relative dielectric constant of Mylar $\varepsilon_r = 5$, of polyurethane $\varepsilon_r = 7$). Figure 3.17

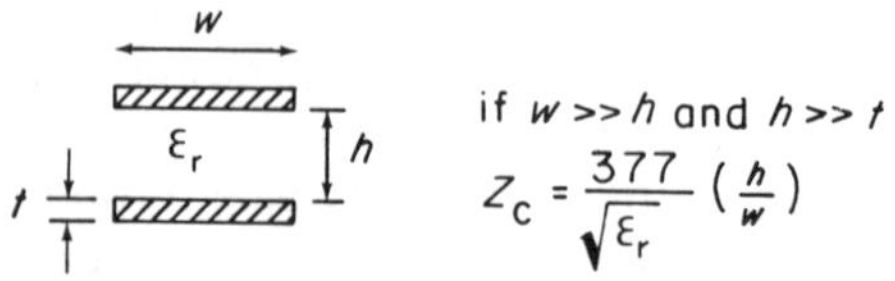

if $w \gg h$ and $h \gg t$

$$Z_c = \frac{377}{\sqrt{\varepsilon_r}} \left(\frac{h}{w}\right)$$

Figure 3.17 Characteristic impedance of parallel flat conductors connection

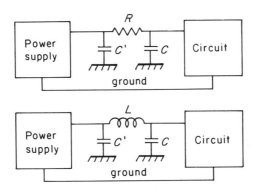

Figure 3.18 RC and LC decoupling filter

shows the parallel flat conductor configuration frequently used for integrated circuits on printed boards.

Most measuring systems have different power supply voltages. Good wiring practices are essential for best performance. Care should be taken to ensure that no digital return signals are present in a path serving as an analogue ground return. Signals are present in a path serving as an analogue ground return. Since the power supplies are not ideal voltage sources, it is necessary to provide some decoupling filters at each circuit. The LC filter is very efficient at high frequencies, but care must be taken to avoid the resonance frequency $f_r = 1/(2\pi\sqrt{(LC)})$.

The RC filter is the usual practice, but the voltage drop in the resistor R reduces the supply voltage. A second capacitor C' can be added to avoid the noise being fed back to the power supply from the circuit.

Generally, the usual +5 V and ±15 V power lines are internally bypassed, but in the system design, it is recommended that additional capacitors be added externally. These capacitors should be located as near the module pins as possible. They would typically be 10 μF (or greater) tantalum types. When the device has an analogue ground pin and a digital ground pin, bypass capacitors should be connected between supply pins and ground pins.

It is important to note that digital ground lines are usually very noisy, and

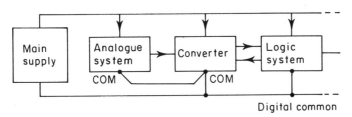

Figure 3.19 Connection between analogue ground and digital ground

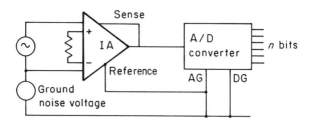

Figure 3.20 Ground voltage is rejected by instrumentation amplifier

have large current spikes. All analogue common leads should be run separately from the digital common leads and tied together at only one point (Figure 3.19).

In some cases, it is possible to use instrumentation amplifiers (I.A.) for converting ground noise voltage into a common-mode voltage which is rejected by the amplifier's differential input (Figure 3.20).

Chapter 4
AMPLIFIERS

The price of operational amplifiers has been reduced considerably during the last ten years while their performances has steadily improved. Nowadays, the op amp gives system designers a useful component to incorporate into numerous applications for realizing many diverse and complex functions. Usually, high performance and flexibility are required for measurement amplifiers, and these are presently realized by combining additional circuits with a basic op amp on the same monolithic chip. In this chapter, the special amplifiers used in measurement systems will be presented with a discussion of their principles of operation and the main precautions required for their implementation.

4.1 INSTRUMENTATION AMPLIFIER

An instrumentation amplifier is a closed-loop, differential input gain block. It is a dedicated circuit with the primary function of accurately amplifying the voltage applied to its inputs. Ideally, the instrumentation amplifier responds only to the difference between the two input signals, and exhibits extremely high impedances between the two input terminals, and from each terminal to ground. The output voltage is developed single-ended with respect to ground and is equal to the product of amplifier gain and the difference of the two input voltages. This difference is often called the differential input voltage.

The properties of this model may be summarized as high input impedance, low offset and drift, low nonlinearity, stable gain, high common-mode rejection, and low effective output impedance. Therefore, it is commonly used for applications which capitalize on these advantages. Examples include: amplification for various types of transducers such as strain-gauge deflection bridges, load cells, thermistor networks, thermocouples, current shunts and so forth, preamplification of small differential signals superimposed on high common-mode voltages, etc.

Note that an instrumentation amplifier differs fundamentally from an operational amplifier. Unlike the instrumentation amplifier, an op amp is an open-loop uncommitted device whose closed-loop performances depend on

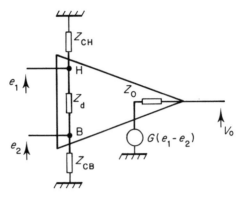

Figure 4.1 Simplified model of an instrumentation amplifier

the external networks used to close the loop. While an op amp can be used to get the same basic transfer function as an instrumentation amplifier, it is generally difficult, and usually impossible, to achieve the same level of performances.

4.1.1 Specifications

4.1.1.1 Gain:

The idealized transfer function is given by:

$$V_0 = G(e_1 - e_2).$$

Where the amplifier gain G is normally set by the user with a single external resistor (from 1 to 1000). Gain temperature coefficient and its nonlinearity can be found on data sheets provided by the manufacturers. The gain linearity is possibly of more importance than the gain accuracy, since the value of the gain can be adjusted to compensate for simple gain errors. The nonlinearity is specified to be the peak deviation from a 'best fit' straight line, expressed as a percentage of peak-to-peak full-scale output.

Note that for the programmable gain instrumentation amplifiers, commonly used in data-acquisition systems, the gain control is accomplished through a number of digital inputs. For example, the Burr Brown 3606 Model programmable gain amplifiers will be discussed in Section 4.5).

4.1.1.2 Offset voltage and input bias currents

Voltage offset and input bias current are often considered as the key figures of merit for instrumentation amplifiers (IA).

Offset voltage is defined as the voltage required at the input to drive the

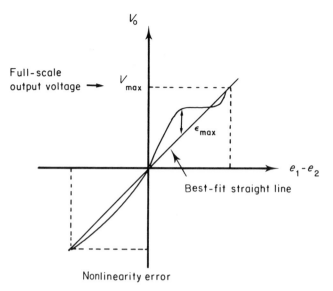

Figure 4.2 Nonlinearity error: $\dfrac{\varepsilon_{\max}}{V_{\max}}$ %

output to zero. While initial offset can be adjusted to zero, shifts in offset voltage with temperature introduce errors. They are by far the most important source of error in most precision measurements. The temperature coefficients of these parameters must be specified by IA manufacturers (example: the input offset voltage drift of AMP 01 Precision IA from "Precision Monolithics Incorporated" is equal to $1.0\,\mu\text{V}/°\text{C}$ in the temperature range $-25°\text{C} \leq T_A \leq +85°\text{C}$).

Offset at the output of an IA consists of two terms:

output offset = V_{OOS} (unity gain output offset)
    + GV_{IOS} (input offset multiplied by the gain G)

The input bias currents are currents flowing into (or out of) the two inputs of the amplifier. They are the base currents (I_b) for bipolar input transistors and the JFET leakage currents (I_G) for the JFET input stage. The bias currents flowing through the source resistance will generate a voltage offset. Initial bias currents are often adjustable to zero but their drift with temperature is very important (JFET leakage current can double every 11°C).

4.1.1.3 Frequency response

Specifications such as bandwidth, slew rate, settling time are common with operational amplifiers.

4.1.1.4 Output impedance

IA output impedance is very low ($\approx 10^{-2}\,\Omega$). It allows very easy and direct interfacing with other following modules. It is very interesting to mention that open-loop output resistance of an op amp is much larger than the output impedance of an IA.

4.1.1.5 Common-mode rejection ratio

An ideal IA must respond only to the difference between the input voltages. If the input voltages are equal ($e_1 = e_2 = V_{CM}$, the common-mode voltage), the output of the ideal IA will be zero.

However, the output voltage of a realistic IA has two components: the first component is proportional to the differential input voltage $e_d = e_1 - e_2$; the second component is proportional to the common-mode input voltage. The common-mode voltage which appears at the amplifier's input terminal is defined as

$$V_{CM} = \frac{e_1 + e_2}{2}.$$

Figures 4.3 and 4.4 illustrate two examples of common-mode interference. In Figure 4.3 it is required to measure the e.m.f. e developed across the thermocouple which is bonded to a metal plate which is itself at a potential V_{CM}. In Figure 4.4 it is required to measure the mechanical stress with a strain-gauge bridge. When the applied stress is zero, the bridge is at balance and the e.m.f. e is equal to zero, but the common-mode voltage, which is equal to

$$V_{CM} = \frac{V_A + V_B}{2} \approx \frac{E}{2},$$

is applied at the amplifier's input terminal.

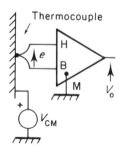

Figure 4.3 Thermocouple bonded to a metal plate which is at a potential V_{CM}

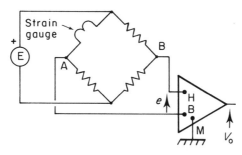

Figure 4.4 Measurement of mechanical stress with a strain-gauge bridge

Common-mode interference is essentially important in data-acquisition systems such as industrial-process control systems whose remote transducers are often spread over miles from measurement instruments. Analogue data must be transmitted over long distances. Since the input lead resistances are no longer negligible, undesirable voltages can be developed and added to any common-mode signal produced by transducers and potential difference between two ground connections.

Figure 4.5 shows an equivalent circuit of the measurement system. R_1 and R_2 represent the input lead resistances, Z_d the IA input impedance, and Z_{CH} and Z_{CB} the leakage impedances from H to ground and B to ground respectively. Considering the common-mode voltage V_{CM} only, this circuit may be drawn as shown in Figure 4.6. In this bridge circuit, it is easy to note that the undesirable voltage V_p produced by V_{CM} is zero only when the bridge is perfectly balanced. When specifying the error caused by V_{CM}, it is usual to consider the worst case unbalance.

Assume that the IA used in this measurement system is ideal, that is to say that its own CMRR is infinite.

The IA measures the potential difference V_p which exists across the H and

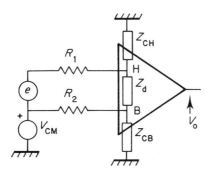

Figure 4.5 Equivalent circuit of measuring system

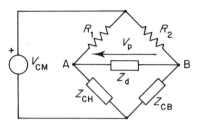

Figure 4.6 Parasitic voltage V_p produced by common mode voltage V_{CM}

B input terminals. The CMRR of the measurement system is expressed in dB by

$$\text{CMRR} = 20 \log_{10} \frac{V_{CM}}{V_p}.$$

The worst case unbalance usually specified by measurement instrument manufacturers is that R_1 is zero and R_2 is some finite resistance (for example 1 kΩ).

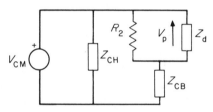

Figure 4.7 Equivalent circuit in the worst case unbalance ($R_1 = 0$)

The equivalent circuit is now shown in Figure 4.7, thus the CMRR is expressed by:

$$\text{CMRR} = 20 \log_{10} \frac{Z_{CB} + R_2 // Z_d}{R_2 // Z_d}.$$

Normally $Z_d \gg R_2$ and $Z_{CB} \gg R_2$, so we can write

$$\text{CMRR} = 20 \log_{10} \frac{Z_{CB}}{R_2}.$$

Numerical example
(1) V_{CM} is a d.c. voltage:
Leakage resistance between B and ground: $Z_{CB} = R_{CB} = 100$ MΩ, lead unbalanced resistance: $R_2 = 1000$ Ω

$$\text{CMRR} = 20 \log_{10} \frac{10^8}{10^3} = 100 \text{ dB}.$$

(2) V_{CM} is an a.c. voltage:
The effects of stray capacitance are no longer negligible. Z_{CB} is effectively a capacitive impedance:

$$Z_{CB} = R_{CB}//100 \text{ pF}$$

At 50 Hz, Z_{CB} is approximately equal to 23 MΩ. Thus at 50 Hz we have

$$\text{CMRR} = 20 \log_{10} \frac{23 \times 10^6}{10^3} \approx 87 \text{ dB}.$$

It is essential to improve the CMRR, especially in precision low-level measurements. Guard techniques allow effective shunting of the common-mode bridge, and at the same time they provide high common-mode impedances (Z_{CH} and Z_{CB}).

The input circuit of the IA are mounted inside a guard shield which must be well isolated from ground and connected directly to the common-mode source V_{CM} as shown in Figure 4.8. Note that R_G must be as low as possible and Z_l represents the leakage impedance from the guard shield to ground. Referring to Figure 4.9, we can write

$$V_p = V_{CM} \frac{R_G}{R_G + Z_l} \frac{R_2}{Z_{CB}} \approx V_{CM} \frac{R_G}{Z_l} \frac{R_2}{Z_{CB}}$$

for $R_G \ll Z_l$

$$\text{CMRR} = 20 \log_{10} \frac{Z_{CB}}{R_2} \frac{Z_l}{R_G}$$

Numerical example

$R_2 = 10^3 \Omega, \quad R_G = 100 \, \Omega, \quad Z_{CB} = 100 \, \text{M}\Omega, \quad Z_l = 1000 \, \text{M}\Omega.$

$$\text{CMRR} = 20 \log_{10} 10^5 + 20 \log_{10} 10^7 = 240 \text{ dB}.$$

When V_{CM} is an a.c. voltage, the effects of stray capacitance must be taken into account.

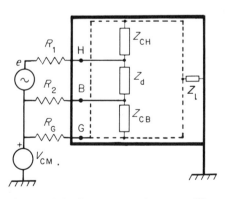

Figure 4.8 Instrumentation amplifier mounted inside a guard shield

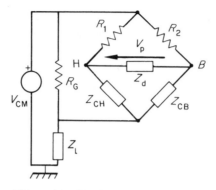

Figure 4.9 Equivalent circuit of measuring system with guard shield

4.1.2 Industrial Realizations

Two basic circuits are employed for monolithic IA realizations. The first circuit approach consists of a differential stage followed by a subtractor stage (Figure 4.10). The two input op amp (A_1 and A_2) are connected in non-inverting configuration to provide high input impedance ($>10^9\,\Omega$). Op amp A_3 is connected in a unity gain differential amplifier.

The circuit has the advantage that the differential gain can be simply adjusted by a single resistor R. It can be seen in Figure 4.10 that

$$V_0 = \left(1 + \frac{2R_1}{R}\right)(e_2 - e_1).$$

Most monolithic IAs include a third stage (A_4) as depicted in Figure 4.11 which adds a great deal of versatility and convenience to the IA. (Example: the 3630 SM Model from Burr Brown includes the two circuits of Figures

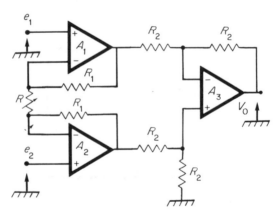

Figure 4.10 Three op. amp Instrumentation Amplifier

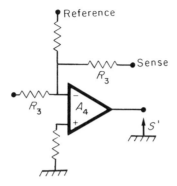

Figure 4.11 Output stage of monolithic IA

4.10 and 4.11). The *sense terminal* is normally connected directly to the output; the *reference terminal* is normally connected to ground.

The reference terminal may be used to offset or to shift the output level to a datum compatible with the load. For instance, where heavy output currents are expected and the load is situated some distance from the IA, voltage drop due to wire resistance will cause error.

The sense terminal may be used to include a booster follower A_p in the feedback loop in order to obtain power amplification without loss of accuracy (Figure 4.13), or to convert the differential input voltage into an output current (Figure 4.12).

The second circuit approach is described by Figure 4.14. It consists of two differential input follower stages which generate the currents:

$$\frac{e_1 - e_2}{R_G} \quad \text{and} \quad \frac{V_0 - E_{\text{ref}}}{R_S}.$$

Amplifiers Z_A and Z_B operate as two current-to-voltage converters. The

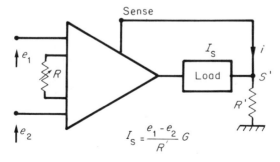

Figure 4.12 Using sense terminal for conversion of differential input voltage into output current

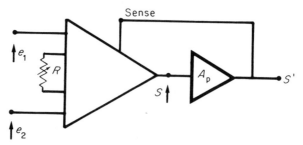

Figure 4.13 Using sense terminal for power amplification

differential input current of Z_A is equal to

$$\Delta i_A = \frac{e_2 - e_1}{R_G} + I_1 - I_2,$$

which is then converted into a voltage V_0. The differential input current of Z_B is equal to

$$\Delta I_B = \frac{V_0 - E_{ref}}{R_S} + I_1 - I_2,$$

which is then converted into a differential voltage. This voltage acts on the current sources in order to compensate the initial unbalanced state produced by the application of $(e_2 - e_1)$. The effect of this operation is to cancel ΔI_A and ΔI_B. Therefore we have

$$I_2 - I_1 = \frac{e_2 - e_1}{R_G} = \frac{V_0 - E_{ref}}{R_S}$$

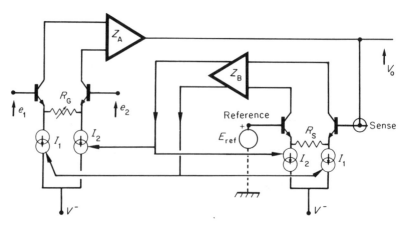

Figure 4.14 Functional diagram of a monolithic IC instrumentation amplifier

and

$$V_0 = E_{ref} + \frac{R_S}{R_G}(e_2 - e_1).$$

The gain of the IA is defined by

$$G = \frac{R_S}{R_G}.$$

It is adjustable by R_G if R_S is constant.

Example of industrial realization An example of a device using the latter concept is the AD 521 J developed by Analog Devices. Its main characteristics are:
Programmable gain: from 0.1 to 1000;
Nonlinearity error: 0.1% max.;
Differential input impedance: $3 \times 10^9 \, \Omega //1.8$ pF;
CMRR, dc to 60 Hz with 1 kΩ source unbalance:
  $G = 1$ CMRR = 70 dB min.
  $G = 100$ CMRR = 100 dB min.;
Output impedance: 0.1 Ω;
Input offset voltage: 3 mV max.;
Voltage drift vs. temperature: 15 μV/°C max.;
Input bias current: 80 nA max.;
Current drift vs. temperature: 1 nA/°C max.

At present, several manufacturers offer many high-accuracy op amp categories, which should not be confused with monolithic IC Instrumentation amplifiers. An op amp is merely a high-gain amplifier requiring the addition of external feedback to be able to operate as a linear amplifier, which cannot offer the user a performance comparable to monolithic IC instrumentation amplifiers.

IC instrumentation amplifiers are especially designed for applications involving the measurement of low-level currents or small voltages from high-impedance sources. Therefore they are commonly employed for high-accuracy interfacing to many categories of transducers. Several high-performance compact signal-conditioning modules include an instrumentation amplifier. Figure 4.15 shows an example of a monolithic signal conditioner which consists of a low-noise and low-drift IA, a low-pass filter and an adjustable transducer excitation. Gain, filter cut-off frequency, output offset voltage and regulated excitation are all adjustable, making the device a very versatile transducer interface module. The low cost and the small size of this high-performance signal conditioner allow the data-acquisition system designer to use a conditioner for each transducer channel. This approach is often needed to reduce noise and to improve resolution.

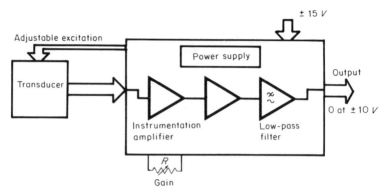

Figure 4.15 High performance compact signal conditioning modulus include IA

Example of industrial realizations Complete signal conditioners: Model 2B 30 and 2B 3I from Analog Devices
Features
  Low drift: 0.5 μV/°C max.;
  Low noise: 1 μV p.p. max.;
  Low nonlinearity: 0.0025% max.;
  High CMR: 140 dB min (60 Hz, G = 1000);
  Adjustable low-pass filter: 60 dB/decade;
  Programmable transducer excitation.
Applications
Measurement and control of pressure, temperature, strain, stress, force, torque.

4.2 ISOLATION AMPLIFIER

The common-mode voltage of an IA is limited to values within the supply voltage range (± 15 V). There are many applications, however, in which the signal to be measured is superimposed on a considerably higher common-mode voltage of perhaps several kilovolts. To deal with such high potentials it is necessary to provide a galvanic isolation between the input circuit and the output of the amplifiers. The isolation amplifier usually consists of an input instrumentation amplifier followed by a unity gain isolation stage. d.c. power supply for the input amplifier must also be isolated from d.c. power supply for the output stage.

The desired isolation characteristics can be achieved with a variety of signal couplers. Two techniques that are currently in widest use because of their low cost and easy implementation are *transformer coupling* and *optical coupling*.

4.2.1 Transformer-coupled Isolation Amplifier

Figure 4.16 shows a simplified diagram of a transformer-coupled isolation amplifier. A d.c. power source provides excitation for a high-frequency pulse generator. Isolation of both signal and power is accomplished with a miniature toroid transformer with multiple windings. The a.c. voltage of the internal generator is converted to a d.c. voltage supply for the input and output stages. The internal generator also provides an a.c. signal which is modulated with the input voltage. The modulated signal is then coupled to two matched demodulators which must produce identical voltages at their outputs. The input stage may be connected in various op amp gain configurations, and the output stage may be connected as a buffer.

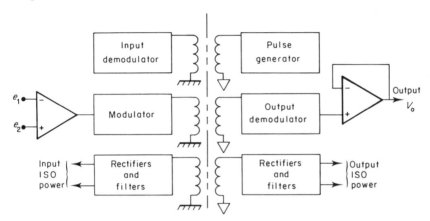

Figure 4.16 Transformer coupled isolation amplifier

Example of industrial realization
Models 3656 AG/3656 BG from Burr Brown.
Features
  Isolation voltage; (continuous d.c.): 2000 V
                            a.c.: 1140 V r.m.s.
                            Pulse test: 5000 V peak.
  Rejection (d.c.): 160 dB.
           60 Hz: 130 dB.
  Leakage current (120 V, 60 Hz): 0.5 μA max.

Note: The leakage current flows between the input common terminal and the output common terminal (across the isolation barrier) with a specified voltage applied across it.

4.2.2 Optically Coupled Isolation Amplifier

The basic operation of an optical isolation amplifier is presented in Figure 4.17. Two matched photo transistors are used, one in the input stage (T_1)

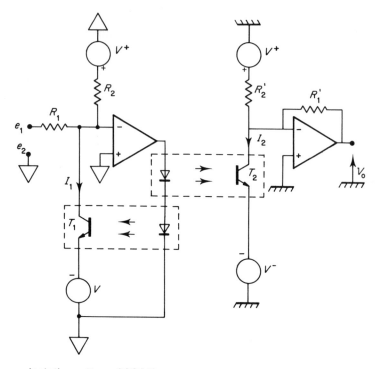

Isolation voltage: 2000 V
Leakage current: 0.25 μA maximum

Figure 4.17 Optically coupled isolation amplifier

and one in the output stage (T_2). In the input stage, the current through the LED (light emitting diode) is controlled by the input op amp. The feedback loop is closed via the reference coupler. In order to permit transmission of bipolar input, it is necessary to superimpose on the useful input current $(e_1 - e_2)/R_1$, a constant current V^+/R_2; referring to Figure 4.17 we can write:

$$I_1 = \frac{e_1 - e_2}{R_1} + \frac{V^+}{R_2}.$$

Since the two opt-couplers are closely matched and since the two photo-transistors receive equal amounts of light from the LED (the two LEDS are excited by the same current), we obtain at the output photo-transistor the current $I_2 = I_1$, the output op amp is connected as a current-to-voltage converter:

$$V_0 = \frac{R_1'}{R_1}(e_1 - e_2) \quad \text{for} \quad R_2 = R_2'.$$

R_1' is an internal one megohm scaling resistor, so the overall transfer function is

$$V_0 = (e_1 - e_2) \frac{10^6}{R_1}$$

with R_1 expressed in ohms.

Example of industrial realization
Models 3650/3652 from Burr Brown.
Features
  Isolation voltage (continuous d.c.): 2000 V
          Pulse test: 5000 V peak;
  Rejection (d.c.): 140 dB
        60 Hz: 120 dB;
  Leakage current (240 V, 60 Hz): 0.25 μA max;
  Leakage capacitance: 1.8 pF;
  Wide bandwidth: 15 kHz, 1.2 V/μs slew rate.

Isolation amplifier performance requirements vary significantly depending on the desired functional characteristics. In applications where bandwidth and speed of response are more important than gain accuracy and linearity, optically coupled isolation amplifiers may be the best choice. For applications where gain accuracy and linearity are essential parameters, a transformer-coupled isolation amplifier may be the suitable choice.

4.2.3 Special Characteristics of Isolation Amplifiers

The following paragraph outlines special characteristics of isolation amplifiers. In the choice of an isolation amplifier, the designer has to understand the significance of special features provided by manufacturers.

Usually, an isolation amplifier, which is represented by the symbol shown in Figure 4.18, consists of the three parts (a) the power supply ($-V$ and $+V$), (b) the input stage with floating input terminals, and (c) the output circuit which are all isolated from one another.

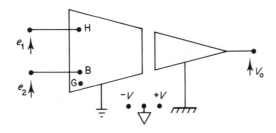

Figure 4.18 Symbolic representation of an isolation amplifier

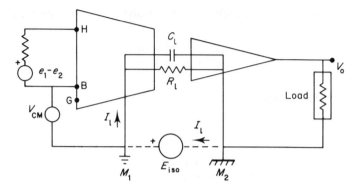

Figure 4.19 Leakage current through an isolation amplifier

4.2.3.1 Common-mode voltage and isolation voltage

When the input common is grounded, the input signal $(e_2 - e_1)$ can be floated by the amount V_{CM} above the input ground (Figure 4.19), V_{CM} is defined as the common-mode voltage (CMV) and is generally limited by the CMV rating of the input stage amplifier (typically ±15 V).

The isolation voltage E_{iso} as shown in Figure 4.19 is the potential difference between the input common and the output common terminals. This feature is very useful when two distinct ground connections are needed. In applications involving a large common-mode voltage, the input common terminal should not be connected to any ground, but the output common terminal has to be connected to the system ground (M_2); in such a case, the term V_{CM} becomes negligible and we can write:

$$E_{CM} = V_{CM} + E_{iso} \approx E_{iso}.$$

E_{CM} is considered as the system common-mode voltage by some manufacturers.

4.2.3.2 Common-mode rejection (CMR) and isolation-mode rejection (IMR)

The above definition of the common-mode voltage and the isolation voltage allows us to understand the difference between the CMR and the IMR.

The CMR is a specification similar to that for instrumentation amplifiers. It indicates the ability to reject the CMV while amplifying only the differential input (referencing signals to the input common).

The IMR indicates the ability to reject the common-mode input signals while transmitting the signals across the isolation barrier (common mode with reference to the output common). The isolation-mode rejection ratio

(IMRR) is defined by the equation:

$$V_0 = G\left(e_d + \frac{V_{CM}}{CMRR}\right) + \frac{E_{iso}}{IMRR}$$

with $e_d = (e_1 - e_2)$

4.2.3.3 Leakage impedance

The impedance of the isolation barrier between the input common terminal and the output common terminal is not infinite and so a leakage current can flow through this impedance, which is usually defined by R_l in parallel with C_l. (Example: the model 3455 from Burr Brown presents a leakage impedance equal to $10^{12}\,\Omega//16\,\text{pF}$).

4.2.4 Applications of Isolation Amplifiers

In those many applications where instrumentation amplifiers cannot provide an adequate solution, isolation amplifiers should be used. Some of these situations may be:

— Measuring low-level signals in the presence of high common-mode voltages (CMV ≫ 15 V).
— Eliminating source ground connections such as in amplifying from floating input sources such as thermocouples.
— Providing an interface between patient monitoring equipment and transducers which are in contact with the patient. In such applications high CMR is required to recover patient signals in the presence of high CMV.
— Providing isolation protection in case of fault conditions.

Examples of applications (1) An application in biomedical designs is illustrated by Figure 4.20 where the isolation amplifier is used to extract the foetal heart beat masked by the mother's heart beat, residual electrode voltages and 50 Hz power-line

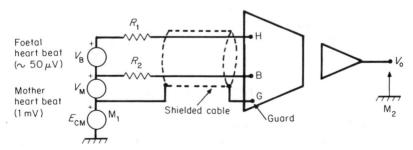

Figure 4.20 Foetal heartbeat monitoring

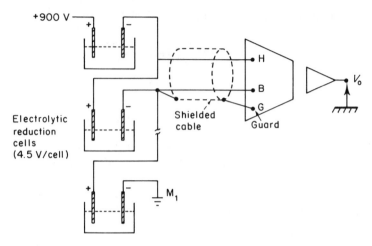

Figure 4.21 Protection of process control circuitry

pick up. It is interesting to note that:
- The CMR from input to output ground allows screening out of the 50 Hz pick up (E_{CM}) and external interferences.
- And the CMR from input to guard allows elimination of the mother's heart signals (typically $V_M = 1$ mV $\approx 200\, V_B$).

(2) An application in industrial control is given in Figure 4.21 where the isolation amplifier is used to protect the process control circuitry from eventual faults in an aluminium production systems. The isolation amplifier must have very high CMV rating ($\gg 900$ V).

4.3 CHARGE AMPLIFIER

4.3.1 Necessity for Charge Amplifier

Asymmetrical crystalline materials such as quartz and barium titanite produce an electrical charge when weights are placed on them. This property is used in piezo-electric transducers which are intended for dynamic force measurement. The electrical charge Q produced is proportional to the applied force F. Therefore the piezo-electric transducer can be modelled as a current source

$$i\left(=\frac{dQ}{dt}\right) \text{ in parallel with a capacitor } C_c.$$

In practical measurement systems, transducers are often remote from measurement amplifiers. The connection between transducers and amplifiers is realized with coaxial cables which present a capacitance C depending on the cable length (Figure 4.22).

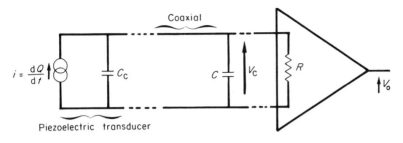

Figure 4.22 Capacitor C depends on the cable length

Assume that:

$$L[Q(t)] = Q(p) \quad \text{(Laplace transform)}.$$

Referring to Figure 4.22 we can write:

$$V_c(p) = \frac{pQ(p)R}{1 + R(C_c + C)p}, \quad (4.1)$$

where R is the amplifier input resistance. The drawback of this measurement system is that:

— The output voltage of the transducer depends on the cable length, that is to say that a new calibration has to be carried out every time the cable length must be changed.
— The amplifier operates as an active high-pass filter, the cut-off frequency of which is expressed by

$$f_c = \frac{1}{2\pi R(C_c + C)}. \quad (4.2)$$

As the frequency response at low frequencies is restricted slowly varying phenomena cannot be measured. Thus a charge amplifier is needed to interface the piezoelectric transducer.

4.3.2 Principle of Operation of a Charge Amplifier

A simplified diagram of a charge amplifier is illustrated in Figure 4.23. Assuming that amplifier A is ideal (infinite gain, infinite input resistance) then the electrical charge Q produced by the transducer will flow through the capacitor C_1. Therefore the amplifier output voltage will be equal to

$$V_0 = -\frac{Q}{C_1},$$

as this is no longer dependent on C and R, recalibration is not necessary when the cable length is changed.

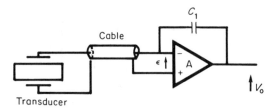

Figure 4.23 Functional diagram of a charge amplifier

4.3.3 Imperfections of Actual Charge Amplifiers

Taking into consideration the finite values of gain A and input resistance R, the amplifier output signal can be expressed as

$$V_0(p) = -Q(p) \frac{ARp}{1 + R[C_c + C + C_1(1+A)]p}. \tag{4.3}$$

It is as if a high-value capacitor $C_1(1+A)$ due to Miller effect were placed in parallel with the cable and the transducer capacitor. Since $C_1(1+A) \gg C_c + C$, the output signal is effectively independent of the cable length, whilst the cut-off frequency, which is equal to

$$\frac{1}{2\pi R C_1(1+A)},$$

is now distinctly less than

$$\frac{1}{2\pi R(C_c + C)}$$

as in Eq. (4.2).

The high input resistance R implies the need for an op amp with a low input bias current. This need can be met through the use of FET input op amps, or by using bipolar input op amps specifically designed for low input bias currents. FET input op amps have well under 1 nA input bias current at 25°C, but the current rises with increasing temperature to values above the bipolar input types. The charge amplifier designer must take care to choose an adequate op amp because an output voltage increasing with time is developed by this small bias current (i_b):

$$V_0 = -\frac{i_b}{C_1} t. \tag{4.4}$$

It is possible to avoid this integration effect by placing a large resistance R_1 (1 to 10 MΩ) in parallel with C_1, but it will cause an output offset voltage:

$$\Delta V_0 = -R_1 i_b \tag{4.5}$$

4.4 CHOPPER AMPLIFIER

To avoid the drift problems usually associated with direct coupled amplifiers, chopper amplifiers are often used for microvolt-range measurement. In a chopper amplifier the direct input voltage is converted into an alternating voltage, amplified by an a.c. amplifier and then converted back into a d.c. voltage proportional to the original input signal.

The principle of operation of a chopper amplifier is illustrated in Figure 4.24.

The two analogue switches sw1 and sw2 are controlled by the same signal, at frequency f_0, generated by the oscillator. They are used as choppers for pulsed modulation (conversion from d.c. into a.c.) and demodulation (conversion from a.c. back into d.c.).

The input to the amplifier is a square-wave carrier voltage with an amplitude proportional to the level of the input voltage and a frequency equal to the oscillator frequency (Figure 4.25). Referring to Figure 4.25, we can write:

$$e_1(t) = e(t)\left[\frac{1}{2} + \frac{2}{\pi}\cos \omega_0 t - \frac{2}{3\pi}\cos 3\omega_0 t + \frac{2}{5\pi}\cos 5\omega_0 t + \cdots\right]. \quad (4.6)$$

Assume that:

(1) V_{os} is the input offset voltage and ΔV_{os} is the input offset voltage drift with temperature of op amp A_1.
(2) The bandwidth of A_1 (gain $= -R_2/R_1$) is narrow enough to eliminate all the harmonics of $e_1(t)$.

At the output of A_1 we obtain

$$e_2(t) = -\frac{R_2}{R_1}e(t)\left(\frac{1}{2} + \frac{2}{\pi}\cos \omega_0 t\right) + \left(1 + \frac{R_2}{R_1}\right)V_{os} + \left(1 + \frac{R_2}{R_1}\right)\Delta V_{os}. \quad (4.7)$$

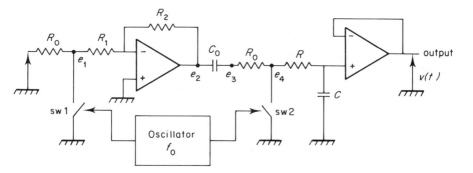

Figure 4.24 Principle of operation of a chopper amplifier

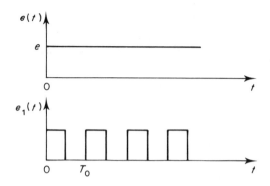

Figure 4.25 The d.c. input voltage is converted into an a.c. voltage $\left(\text{frequency } f_0 = \dfrac{1}{T_0}\right)$

Equation (4.7) indicates that V_{os} and ΔV_{os} are amplified by A_1. The direct component of $e_2(t)$ is then eliminated by the capacitor C_0. The demodulation action is achieved by the analogue switch sw2 which provides

$$e_4(t) = e_3(t)\left[\frac{1}{2} + \frac{2}{\pi}\cos \omega_0 t - \frac{2}{3\pi}\cos 3\omega_0 t + \cdots\right] \quad (4.8)$$

with

$$e_3(t) = -e(t)\frac{R_2}{R_1}\left(\frac{2}{\pi}\cos \omega_0 t\right)$$

The low-pass filter (RC) allows preservation of only the direct component of $e_4(t)$ so that the output signal is equal to

$$v(t) = -e(t)\frac{R_2}{R_1}\frac{4}{\pi^2} + V_{os2} \quad (4.9)$$

The offset voltage V_{os2} of the follower A_2 may be either negligible or turned out using an offset nulling potentiometer.

Numerical example For the purpose of digitizing a signal produced by a thermometer with a typical 1 mV, variation in the frequency band =0.1 Hz, a chopper amplifier has to be designed with a gain =1000 in order to minimize the offset voltage influence of op amp (LF 356 drift = 5 μV/°C).
  Operating ambient temperature range $\Delta T° = 40°C$.
  Thermometer sensitivity: 40 mV/°C.
Solution. Since LF 356 is a JFET input op amp, its input bias current may be negligible. If the signal amplification is achieved by a standard amplifier as shown by Figure 4.26, the input offset voltage drift will cause at the output an error equal to

$$\Delta v = 5 \,\mu\text{V} \times 40°\text{C} \times 1000 = 200 \,\text{mV},$$

which corresponds to an error of 5°C on the temperature measurement.

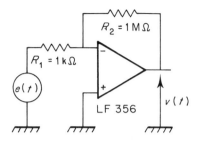

Figure 4.26 Errors caused by offset voltage drift are not acceptable

The error due to $(1 + R_2/R_1)V_{os}$ is not troublesome because it is constant and can be easily compensated.

Referring to Figure 4.25 and Eq. (4.9), we note that the chopper amplifier allows reduction of the error caused by the offset voltage drift from

$$200 \text{ mV} \quad \text{to} \quad V_{os2} = 5\,\mu\text{V} \times 40 = 200\,\mu\text{V},$$

which corresponds to an error of $5 \times 10^{-3}\,°\text{C}$ on the temperature measurement.

Analogue switches can be realized by using JFETs driven by square waves V_c as shown in Figure 4.27.

— When $V_c = 0$, the JFET is ON. Since $R_{DS} \ll R_0$, we have $e_1 = 0$
— When $V_c = -V_0$ ($V_0 > V_p$), the JFET is OFF. We obtain $e_1(t) = e(t)$.

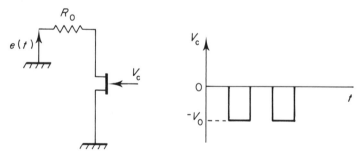

Figure 4.27 Analogue switch operation

Example of industrial realization
Model 235 from Analog Devices.
Features
    Input offset drift: $0.1\,\mu\text{V}/°\text{C}$.
    Long-term stability: $5\,\mu\text{V}/\text{year}$.
    Low noise: $0.5\,\mu\text{V}$ peak to peak, in the frequency band from 0.01 Hz to 1 Hz.
Applications:
    Microvolt voltage measurements
    Picoampere current measurements
    Precision integrator
    Transducers interfacing.

4.5 PROGRAMMABLE GAIN AMPLIFIERS

In a data-acquisition system, the multiplexer is generally followed by an amplifier and an ADC. The amplifier must have the ability to preserve the signals from the multiplexer. The input signals to the multiplexer are derived from a variety of transducers, such as thermocouples and others, with a very wide range of signal levels possible which require a *programmable gain data amplifier* (PGDA). The PGDA is an amplifier whose gain is controlled by the application of digital signals. Amplifier gain may be *preprogrammed* on a per-channel basis with a knowledge of the sensor characteristics for each channel or it may be adjusted to bring the output signal within a desired range after channel selection.

The circuit of Figure 4.28 shows an instrumentation amplifier converted to a PGDA. The switches operate in pairs to change the gain, usually in steps such as 1–10–100 or 1–16–256.

When a wide range of gains is required with fine resolution it is possible to add a second stage as illustrated in Figure 4.29. Assume that the first stage provides gain of 1–10–100, and the second stage provides gains from 0 to 15. Thus the overall amplifier can provide 46 gain positions from 0 to 1500.

The most important specification of a PGDA is undoubtedly its speed of response. When a wide range of gains is necessary, and speed is also important, it is usually recommended to cascade two PGDAs, because the

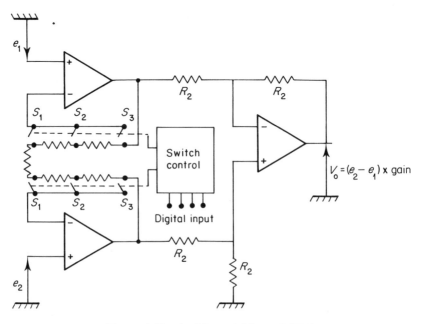

Figure 4.28 A differential-input PGDA

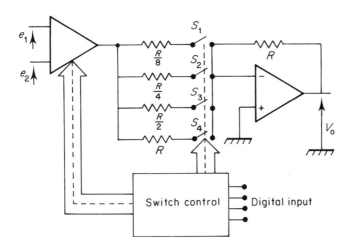

Figure 4.29 A two-stage differential PGDA

overall gain–bandwidth product is greater than for a single PGDA operating over the same range of gains. It is useful to remember that the response speed of the amplifier decreases as gain is increased.

4.6 VOLTAGE TO CURRENT CONVERTER USING INSTRUMENTATION AMPLIFIERS

In industrial instrumentation and control systems, it is frequently necessary to produce a current output signal which is proportional to the voltage input.

The design of a voltage-to-current converter consists of an instrumentation amplifier (IA), precision resistors and a high-stability voltage reference. Figure 4.30 shows an example of this class of converter. Assume that

$$V_0 = G(e_2 - e_1), \tag{4.10}$$

where G is the IA gain and $(e_2 - e_1)$ the input voltage.

The current through the load resistance is

$$I_L = \frac{S'}{R} + I_1$$

If $I_1 \ll I_L$, i.e. if $R_3 \gg R_L$, then

$$I_L = \frac{G}{R}(e_2 - e_1). \tag{4.11}$$

I_L is therefore independent of the voltage across the feedback resistor and the load resistance R_L. Such a load is called *off ground* or *floating*. For this kind of operation, the load may in practice consist only of passive elements;

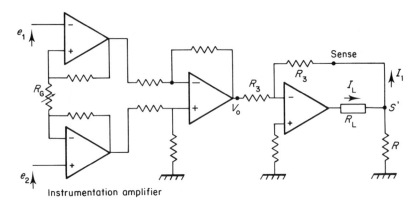

Figure 4.30 Voltage to floating current converter

as for active loads, there is normally a connection to ground. Voltage-to-current converter for grounded loads may be supplied by the circuit of Figure 4.31.

The output voltage of the IA is

$$V_0 = G(e_2 - e_1) + V_{\text{ref}}, \qquad (4.12)$$

which can be also expressed by

$$V_0 = RI_L + V_{\text{ref}};$$

then

$$I_L = \frac{G}{R}(e_2 - e_1).$$

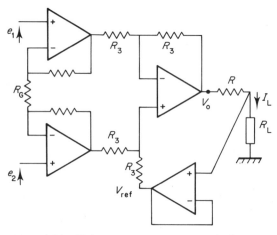

Figure 4.31 Voltage to current converter for grounded loads

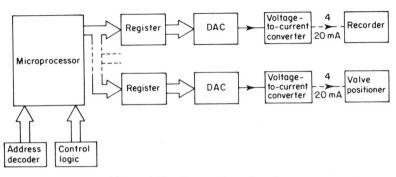

Figure 4.32 Current loop interface

The nominal output current range of an industrial voltage-to-current converter is usually 4–20 mA. In a typical application this device may act as an interface between the DAC output and a process control element such as a variable position actuator. It is also designed for a transmission link between subsystems and elements of process control systems such as recorders, actuators, indicators, motors, etc. Figure 4.32 shows an example of application in a multichannel microprocessor system. An analogue output board with 4–20 mA outputs is often necessary to drive final control elements.

The transmission of current reduces noise and improves system signal-to-noise ratio, since the information is unaffected by induced voltage noise, stray emfs due to thermocouple effects and voltage drops in the line. Another benefit of this form of transmission is that it permits several loads to be connected in series up to a specified maximum voltage using a V/I converter as a transmitter.

Thus only two wires are needed for transmission. We note that the offset represented by 4 mA provides a distinction between span or scale zero and no information due to a zero current flow caused by an open circuit.

It is easy to build a voltage-to-current converter using a standard op amp, provided attention is paid to the requirements for accuracy. A small size modular voltage-to-current converter may offer a low-cost system solution with low offset error and low nonlinearity error.

Chapter 5
ANALOGUE SIGNAL PROCESSING FOR MEASUREMENT SIGNALS

5.1 ANALOGUE FILTERS

It is not the intention to recall here the whole of filter design theory. We intend to point out only the practical considerations governing the use of filters in measuring systems.

5.1.1 Preliminaries

When measuring a voltage or a current, both time filtering and frequency filtering are unavoidably applied to the measurement signal.

(1) Time Filtering

The signal is observed only between $t_0 - T_m/2$ and $t_0 + T_m/2$. By examining the 'gate function' $\sqcap_{T_m/2}(t)$ defined as shown in Figure 5.1, it can be seen that the measured signal is equal to

$$x_m(t) = x(t)\sqcap_{T_m/2}(t).$$

The spectrum of this $x_m(t)$ is given by

$$\hat{x}_m(f) = x(f) * \sqcap_{T_m/2}(f) \quad \text{(convolution product)},$$

but the Fourier transform of $\sqcap_{T_m/2}(t)$ is equal to

$$\sqcap_{T_m/2}(f) = T_m \frac{\sin \pi T_m f}{\pi T_m f}(\cos 2\pi t_0 f - j \sin 2\pi t_0 f)$$

The time filtering thus changes the spectrum of the measurement signal.

(2) Frequency Filtering

In a measuring system, the signal carrying the useful information to be measured has to pass through various devices which unfortunately cause frequency filtering such as connecting cable wire, transmitting channels, etc.

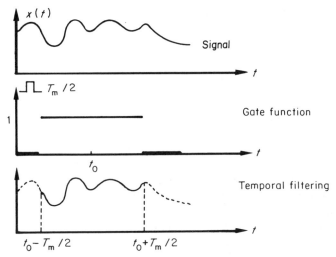

Figure 5.1 The measuring deforms the signal spectrum

The following filters are electronic devices that are deliberately introduced in the measuring system in order to keep only the useful part of the spectrum of the signal.

5.1.2 Passive Filters

These are made using *RL, RC,* or *RLC* type basic cells. From the signal processing point of view such filters will only dissipate signal energy although another drawback is the use of inductance coils. For low frequencies, these components are big, heavy and expensive. Moreover, their resistivity is no longer negligible.

5.1.3 Active Filters

These are now very common and useful due to the low cost and increasing quality of the op amp.

They have the following advantages:

— As the energy dissipated in passive components may be compensated for, it is possible to increase the useful signal energy.
— The cells may be connected to each other without modifying their properties due to loading effects.
— No inductance coils are needed.

Numerous design methods exist. Some do not require many amplifiers but they have a major drawback:

(1) A high sensitivity due to variation of the filter characteristics such as cut-off frequency, Q, band-pass gain with the component values.

(2) Consequently their adjustment is very difficult. To avoid such difficulties industrial devices are modular blocks, including many amplifiers, which are thus able to meet many industrial requirements.

Low-pass, high-pass, band-pass and band-stop filters all exist in integrated, easy to operate blocks. Cut-off frequencies, ranging from 10^{-3} Hz to some 10 kHz, which are controlled by a single external resistor, a voltage, or a digital control, are possible.

The three types of filter responses which are available, are Bessel, Butterworth and Tchebyschev. Tchebyschev filters display a steep cut-off but there is a ripple in the pass band, and the phase variation versus the frequency is not linear. Bessel filters display a smoother cut-off zone, but the phase variation versus the frequency remains as linear as possible, which sometimes may be very useful. Butterworth filters are an average type between Bessel and Tchebyschev. These filters display 2, 4, 6 or 8 poles. Their price ranges from (1986) 10 to 300 US dollars. For example, from Frequency Devices: 704L2B low-pass second-order filter Butterworth 1–10 Hz (47 US dollars); 750L2YA2W low-pass second-order Tchebyschev (1 dB ripple) 10^{-3} to 10^{-2} Hz (200 US dollars).

5.1.4 Universal Filters

These blocks include operational amplifiers that enable realization of second-order filters with external passive components. The state variable approach is employed for designing. On the integrator circuit of Figure 5.2, if $e(t)$ is the sinewave $e(t) = A \sin \omega t$, then the output is equal to

$$s(t) = -\frac{1}{RC}\left(-\frac{A}{\omega}\cos \omega t\right).$$

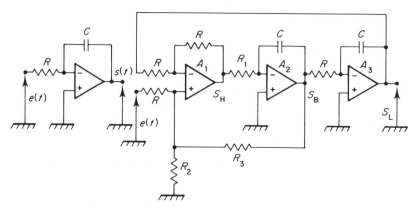

Figure 5.2   Universal filter

We notice that
the output $s(t)$ is proportional to $1/\omega$ and to $e(t)$;
the transmittance depends on the frequency.
The whole filter includes two integrators to create $1/\omega^2$ terms, an amplifier working as a subtraction circuit (Figure 5.2) and a fourth amplifier generally not connected.
Using external components, the designer can make filters of any type of response (Butterworth, Bessel,...) with chosen cut-off frequencies. In Figure 5.2, we have the following transmittances:

$$\frac{s_H}{e} = \frac{-2kRR_1C^2\omega^2}{1+2jk'RC\omega - RR_1C^2\omega^2}$$ corresponding to a high-pass filter

$$\frac{s_B}{e} = \frac{-2k'RC\omega j}{1+2jk'RC\omega - RR_1C^2\omega^2}$$ Corresponding to a band-pass filter

$$\frac{s_L}{e} = \frac{2k}{1+2jk'RC\omega - RR_1C^2\omega^2}$$ corresponding to a low-pass filter

with

$$k = \frac{R_2//R_3}{R_2//R_3 + R} \quad \text{and} \quad k' = \frac{R_2//R}{R_2//R + R_3}.$$

We notice that:

— The cut-off frequency is adjustable with RR_1C^2 term because it is defined by the expression $R_1^2C^2\omega = 1$.
— The gain in the pass band is controlled by k.
— The kind of response of the filter is connected with the damping factor $2jk'RC\omega$ in the denominator of the transmittance. It is adjustable with k'. The value of Q also depends directly on the damping factor and is adjustable with k'.

Example of industrial device
AF 100 circuit from National Semiconductors:
— Upper frequency limit: 10 kHz.
— $Q_{max} = 500$.
— Approximate price: 10 US dollars (1986).

5.1.5 Filters Using Generalized Impedance Converters

This module includes two operational amplifiers. It enables the realization of the structure of Figure 5.3 which displays between A and ground the impedance

$$Z_{AG} = \frac{Z_1 Z_3 Z_5}{Z_2 Z_4}.$$

Both amplifiers are assumed to be ideal ones.

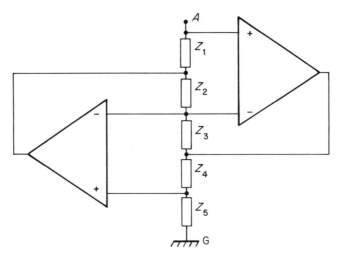

Figure 5.3 Generalized impedance converter

Example of industrial device
AF 120 circuit from National Semiconductor.
In this module, Z_3 and Z_4 are internal and $Z_3 = Z_4 = R$; thus

$$Z_{AG} = \frac{Z_1 Z_5}{Z_2}.$$

This property enables the realization of
(1) An inductance using resistors for Z_1, Z_5 and a capacitor for Z_2. We get then a dipole $Z_{AG} = K\omega$ (K is constant and real).
(2) A capacitor using resistors for Z_1, Z_2 and a capacitor for Z_5.
This case is not interesting for signal processing or instrumentation.
(3) A super capacitor called FDNR (frequency-dependent negative resistor) using capacitors for Z_1 and Z_5 and resistor for Z_2. We get then a dipole $Z_{AG} = K/\omega^2$ (K is constant and real).

This module is really versatile and enables, for example, the realization of LC filters without a real inductance coil, intended for low frequencies in order to take advantage of their low sensitivity.

5.1.6 Rauch Cells Filters

This kind of filter needs very few operational amplifiers associated with RC cells. The design and use of such filters is straightforward. Consider a filter with a transfer function displaying a single pair of complex conjugate poles, and zeroes only at the origin and at infinity. The basic structure is made with five elements as shown in Figure 5.4.

The transfer function of this filter is equal to:

$$F(p) = \frac{S(p)}{E(p)} = \frac{-Y_1 Y_3}{Y_5(Y_1 + Y_2 + Y_3 + Y_4) + Y_3 Y_4}$$

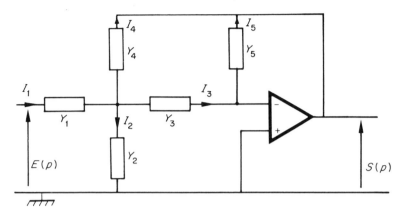

Figure 5.4 Rauch cell filters

Suppose we want to design the low-pass filter with the transfer function

$$F_1(p) = \frac{-K}{ap^2 + bp + c},$$

with a, b, c, and K real.

The filter structure is obtained copying the circuit of Figure 5.4. The values of the components are obtained by identifying both transfer functions: $F(p)$ and $F_1(p)$.

If a high-pass filter is needed, the correct components values are obtained by performing an RC/CR transfonm.

The design of such filters is made using the following values:

— Low-pass: $R_1 = R_2 = R_3 = R$
— High-pass: $C_1 = C_2 = C_3 = C$.

The other components are then calculated using:

— for the low-pass filter: $C_0 = 1/\omega_0 R$ (ω_0 = cut-off angular frequency)
— for the high-pass filter: $R_0 = 1/\omega_0 C$.

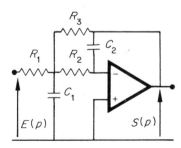

Figure 5.5 Second-order low-pass filter

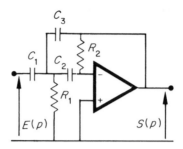

Figure 5.6 Second-order high-pass filter

Table 5.1 gives the different K factors needed to perform the calculation of the other components depending on whether the filter to be designed is Bessel, Butterworth or Tchebyschev type.

The following circuits are third-order filters. The association of second- and third-order filters permits the design of filters of any order.

The filter components have the following values:

Low pass	High pass
$C_1 = K_1 C_0$	$R_1 = R_0/K_1$
$C_2 = K_2 C_0$	$R_2 = R_0/K_2$
$(C_3 = K_3 C_0)$	$(R_3 = R_0/K_3)$

Band-pass and band-notch filters are realized using the schemes shown in Figures 5.9 and 5.10.

Notice. A second-order band-pass filter can be made using the circuit in Figure 5.11.

Its Laplace transfer function is equal to

$$F(p) = \frac{-pG_1 C_3}{p^2 C_3 C_4 + pG_5(C_3 + C_4) + G_5(G_1 + G_2)}.$$

5.1.7 Switched Capacitor Filters

5.1.7.1 Principle of operation

The adjustment of the cut-off frequency of a filter may be realized using a potentiometer or variable capacitor. But in order to suppress manual

Table 5.1 *K* factors

Number of poles	K_1	K_2	K_3	K_4	K_5	K_6	K_7
1	1.00						
2	1.00	0.33					
3	1.19	0.69	0.16				
4	0.51	0.21	0.71	0.12			
5	0.76	0.39	0.12	0.64	0.085		
6	0.35	0.15	0.40	0.12	0.59	0.063	
7	0.71	0.25	0.085	0.37	0.093	0.56	0.049

Bessel

Number of poles	K_1	K_2	K_3	K_4	K_5	K_6	K_7
1	1.00						
2	2.12	0.47					
3	2.37	2.59	0.32				
4	3.19	0.25	1.62	0.61			
5	2.16	4.31	0.21	1.85	0.54		
6	5.79	0.17	2.12	0.47	1.55	0.64	
7	2.10	6.05	0.15	2.40	0.41	1.66	0.60

Butterworth

Number of poles	K_1	K_2	K_3	K_4	K_5	K_6	K_7
1	2.86						
2	2.10	0.31					
3	3.37	4.54	0.18				
4	2.55	0.10	3.54	0.79			
5	5.58	13.14	0.072	5.11	0.41		
6	19.31	0.05	7.07	0.24	5.17	1.23	
7	7.84	26.03	0.03	9.89	0.15	6.50	0.60

Tchebyschev $\pm \frac{1}{2}$ dB

Number of poles	K_1	K_2	K_3	K_4	K_5	K_6	K_7
1	1.96						
2	2.73	0.33					
3	4.21	5.84	0.16				
4	10.75	0.094	4.45	0.80			
5	6.96	15.56	0.06	6.40	0.36		
6	24.12	0.041	8.82	0.30	6.46	1.24	
7	9.77	32.50	0.03	11.70	0.13	8.10	0.55

Tchebyshev ±1 dB

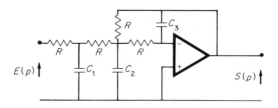

Figure 5.7　Third-order low-pass filter

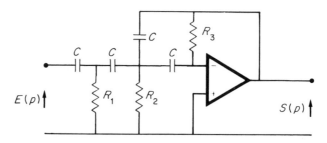

Figure 5.8　Third-order high-pass filter

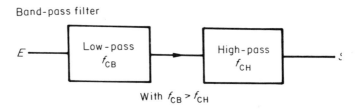

Figure 5.9　Band-pass filter

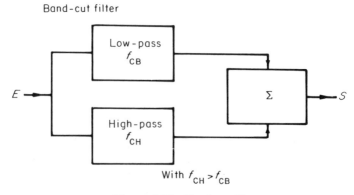

Figure 5.10　Band-cut filter

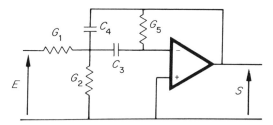

Figure 5.11 Second-order band pass filter

adjustments we can use:

— R_{DS} resistor of a JFET controlled by voltage V_{GS}, but the range is limited.
— a DAC working as a two-quadrant multiplier. The main drawback is the important offset voltage that varies with the digital control.

Switched capacitor technology, which offers a new solution, is based upon the main idea of simulating a frequency-controlled resistor with a capacitor. Assume Figure 5.12 that $V_1 > V_2$. When the switch is connected to 1 the capacitor charges to $Q_1 = CV_1$, whereas with the switch on 2 the capacitor is charged to $Q_2 = CV_2$.

If T_s is the period of the control signal, we will have the charge transfer during T_s:

$$\Delta Q = C(V_1 - V_2).$$

As this is equivalent to the current

$$I = \frac{\Delta Q}{T_s} = \frac{C(V_1 - V_2)}{T_s}.$$

Then the circuit, between terminals A and B, is equivalent to a resistor with the value

$$R = \frac{V_1 - V_2}{I} = \frac{T_s}{C} = \frac{1}{CF_s}.$$

Resistor R is thus simulated by switching a capacitor C at the control

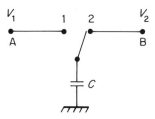

Figure 5.12 Switched capacitor

frequency F_s which is called the *sampling frequency*. This procedure enables the design of filters using switched capacitors. The following paragraphs describe such filters.

5.1.7.2 Basic devices

(1) Basic Low-Pass Filter

In Figure 5.13 circuit signals ϕ_1 and ϕ_2 control switches 1 and 2.

The transfer function $F(p) = S(p)/E(p)$ of the system is determined as follows:

When ϕ_1 goes high, switch 1 is ON and switch 2 is OFF. When $t = kT_s$,

$$Q_{\alpha C}(k, T_s) = \alpha C E(kT_s)$$
$$Q_C = CS(kT_s).$$

When ϕ_2 goes high, switches 1 and 2 change state. And when $t = (k + \frac{1}{2})T_s$, we have:

$$Q_{\alpha C}[(k + \tfrac{1}{2})T_s] = \alpha CS[(k + \tfrac{1}{2})T_s]$$
$$Q_C = CS[(k + \tfrac{1}{2})T_s].$$

When ϕ_1 goes high, t is now $(k + 1)T_s$, so we have

$$Q_{\alpha C}[(k + 1)T_s] = \alpha C E[(k + 1)T_s]$$
$$Q_C = CS[(k + 1)T_s].$$

We notice that ϕ_1 controls the sampling and $\phi_2 = 1$ controls the charge transfer. In each capacitor the charge variation is given by

$$\Delta Q_{\alpha C} = \alpha C\{S[(k + \tfrac{1}{2})T_s] - E(kT_s)\}$$
$$\Delta Q_C = C\{S[(k + \tfrac{1}{2})T_s] - S(kT_s)\}.$$

But the conservation of charges implies that

$$\alpha CS[(k + \tfrac{1}{2})T_s] - \alpha CE(kT_s) + CS[(k + \tfrac{1}{2})T_s] - CS(kT_s) = 0$$

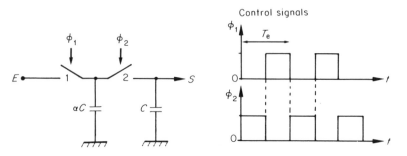

Figure 5.13 Basic low-pass filter

This equation describes the evolution of S for the first half period $T_s/2$. Capacitor C is separated from αC during the second half period. We have then:

$$CS[(k + \tfrac{1}{2})T_s] = CS[(k + 1)T_s]$$

Recall that $S[(k + 1)T_s]$ means the value of S for $t = (k + 1)T_s$.

The last two equations enable elimination of the intermediate variable $S[(k + \tfrac{1}{2})T_s]$:

$$(\alpha + 1)S[(k + 1)T_s] = S(kT_s) + \alpha E(KT_s).$$

The shift theorem for the Z-transform,* which is written as

$$f(t - kT_s) = Z^{-k}F(z).$$

is applied here in order to deduce the Z-transform of both terms of the equation*

$$(\alpha + 1)ZS(Z) = S(Z) + \alpha E(Z).$$

Thus the Z transfer function of the system may be shown to be

$$F(Z) = \frac{S(Z)}{E(Z)} = \frac{\alpha}{\alpha + 1} \cdot \frac{1}{Z - 1/(\alpha + 1)}$$

As Z is approximated by

$$\frac{1}{p} \equiv \frac{T_s}{Z - 1} \rightarrow Z \equiv 1 + pT_s$$

we obtain consequently:

$$F(p) = \frac{1}{T_s} \frac{\alpha}{\alpha + 1} \cdot \frac{1}{p + \dfrac{1}{T_s} \dfrac{\alpha}{\alpha + 1}}.$$

This is the p-expression (Laplace transform) of a first order low-pass filter with the transfer function

$$F(p) = \frac{1}{RC} \frac{1}{p + 1/(RC)}$$

provided the following equivalence applies.

$$\frac{1}{RC} \equiv \frac{1}{T_s} \frac{\alpha}{\alpha + 1}$$

giving

$$RC \equiv T_s \frac{\alpha + 1}{\alpha}.$$

* For the definition of the Z-transform see Section 7.2.5.

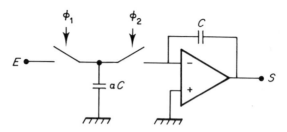

Figure 5.14 Switched capacitor integrator

The $(\alpha + 1)$ term indicates that the charge transfer from αC into C is incomplete.

(2) Switched Capacitor Integrator

Assume that ϕ_1 and ϕ_2 are now used to control the following circuit (Figure 5.14).

The charge conservation for the first half period $T_s/2$ gives the following equation:

$$\alpha CE(kT_s) + C\{S[(k + \tfrac{1}{2})T_s] - S(kT_s)\} = 0. \tag{5.1}$$

Since the capacitor C is separated from αC during the second half period it is evident that

$$CS[(k + \tfrac{1}{2})T_s] = CS[(k + 1)T_s]. \tag{5.2}$$

The Z-transform of this equation gives (using both (5.1) and (5.2)):

$$S(kT_s) - \alpha E(kT_s) = S[(k + 1)T_s]$$

and

$$S(Z) - \alpha E(Z) = ZS(Z).$$

The Z-transfer function of the circuit is

$$F(Z) = \frac{S(Z)}{E(Z)} = \frac{-\alpha}{Z - 1},$$

as before, Z is approximated by $Z = 1 + T_s p$. Hence we obtain

$$F(p) = -\frac{1}{(T_s/\alpha)p}.$$

The expression $F(p) = -1/(RCp)$ is thus recognized as the transfer function of ideal integration.

Notice also that in $RC = T_s/\alpha$ the $\alpha + 1$ term has disappeared. This occurs because the charge transfer is made with a fictitious ground. Depending on the chosen approximation, in this case $Z = 1 + pT_s$, we can have the equivalence shown in Figure 5.15.

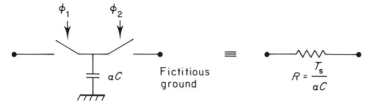

Figure 5.15 A capacitor associated with two switches is equivalent to a resistor

Rough filters can be designed using a classical diagram with resistor, capacitor, operational amplifier and replacing R by switched capacitor systems with $R = T_s/\alpha C$. However for better performance, the following bilinear approximation is commonly used

$$Z = \frac{1 + (T_s/2)p}{1 - (T_s/2)p}$$

The $1 - (T_s/2)p$ term creates a distortion of the filter gain compared to the last approximation. First the gain curve is deformed, second the phase becomes 180° instead of 90° for the usual filter. The circuit of Figure 5.16 reveals also the errors produced by the stray capacitors of analog switches. This problem is solved using the circuit of Figure 5.17 in which αC is a floating capacitor while stray capacitors are discharged to ground at every period.

The basic integrator circuit designed in Figure 5.14 can be replaced by one of Figure 5.18, which integrates using the *delayed-rectangles* method. It

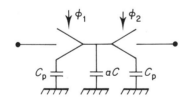

Figure 5.16 Stray capacitors are connected in parallel on αC

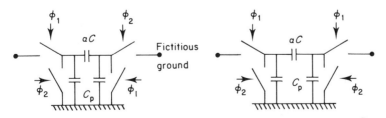

Figure 5.17 Elimination of stray-capacitors effect

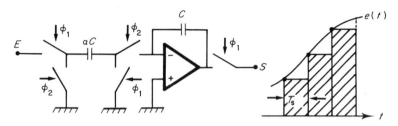

Figure 5.18 Integration using delayed-rectangles method

corresponds to the approximation

$$\frac{1}{p} = \frac{T_s}{Z-1}$$

This rough method doesn't provide good accuracy in the filter transfer function. It explains why the circuit of Figure 5.19 which employs the *Simpson method* is commonly used. The transfer function of this cell, which is

$$\frac{S}{E} = -\frac{\alpha C}{C}\frac{Z+1}{Z-1}$$

corresponds to an integration which realizes the *bi-linear* approximation

$$\frac{1}{p} = \frac{T_s}{2}\frac{Z+1}{Z-1}.$$

As

$$\frac{S}{E} = \frac{-1}{CT_s/(2\alpha C)}\frac{T_s}{2}\frac{Z+1}{Z-1};$$

it is inferred that

$$R = \frac{T_s}{2\alpha C}.$$

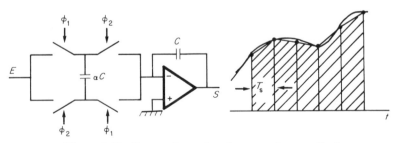

Figure 5.19 Integration using the trapezium method

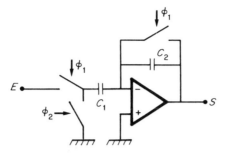

Figure 5.20 Switched-capacitor delay line

The filter synthesis can be realized using the analogue integrator design method (Laplace transfer function). The resistors are simply replaced by four switches and an $\alpha C = T_s/(2R)$ capacitor network.

(3) Switched Capacitor Delay Line

The control signals ϕ_1 and ϕ_2 are defined in Figure 5.13. The delay realized in the circuit of Figure 5.20 is half a period of the sampling signal.

The Z-transfer function is equal to

$$F(Z) = \frac{S}{E} = \frac{C_1}{C_2} Z^{-1/2}.$$

The design of switched capacitor filters can also be made using this delaying block. Some of the most frequently used methods are indicated as follows:

Direct method. Consider the design of the filter having the following transfer function

$$F(Z) = \frac{b_0 + b_1 Z^{-1} + b_2 Z^{-2}}{a_0 + a_1 Z^{-1} + a_2 Z^{-2}}.$$

This can be described by the structure shown in Figure 5.21. It follows that

$$S_1(Z) = \frac{1}{a_0} E(Z) - \frac{a_1}{a_0} Z^{-1} S_1(Z) - \frac{a_2}{a_0} Z^{-2} S_1(Z)$$

and

$$S(Z) = b_0 S_1(Z) + b_1 Z^{-1} S_1(Z) + b_2 Z^{-2} S_1(Z).$$

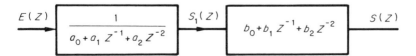

Figure 5.21

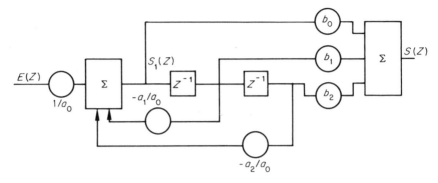

Figure 5.22 Filter structure using direct method

The filter structure is then defined in Figure 5.22.

Cascade method. The transfer function, $F(Z)$, is factorized in order to obtain the form

$$F(Z) = \frac{b_0(Z - Z_1)(Z - Z_2)\cdots}{a_0(Z - Z_3)(Z - Z_4)\cdots}$$

If the root is real, we will have first-order terms. If the roots are conjugate complex, we will have second-order terms such as

$$\frac{(Z - Z_2)(Z - \bar{Z}_2)}{(Z - Z_4)(Z - \bar{Z}_4)},$$

which are the transfer functions of biquadratic cells. The filter structure is defined in Figure 5.23.

Parallel method. In this case the transfer function, $F(Z)$, is split up into first-order fractions

$$F(Z) = \underset{\underset{F_1(Z)}{\downarrow}}{\frac{C_0}{(Z - Z_1)}} + \underset{\underset{F_2(Z)}{\downarrow}}{\frac{C_1}{(Z - Z_2)}} + \underset{\underset{F_3(Z)}{\downarrow}}{\frac{C_3 Z + C_4}{(Z - Z_3)(Z - \bar{Z}_3)}} + \cdots$$

The filter structure is shown in Figure 5.24.

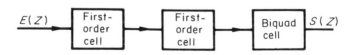

Figure 5.23 Filter using cascade connected cells

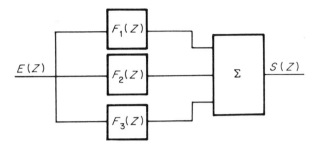

Figure 5.24 Filter structure using parallel connected cells

5.1.7.3 Example of industrial device

The integrated circuit MF 10 (Figure 5.25) is a switched capacitor filter. It is manufactured by National Semiconductor and uses C.MOS technology. It includes two symmetrical circuits, each including

— an input amplifier which, when connected to external resistors, realizes the summing.
— a three-input summing amplifier. One of the inputs can be connected to the analogue ground or the output S_3 to improve the possibilities of the device.
— two clock-controlled integrators. The clock frequency can be divided by 50 (terminal 12 connected to a 5 V supply) or 100 (terminal 12 connected to ground). This permits adjustment of the integrator time constant.

The technology of resistor simulation by a switched capacitor appeared

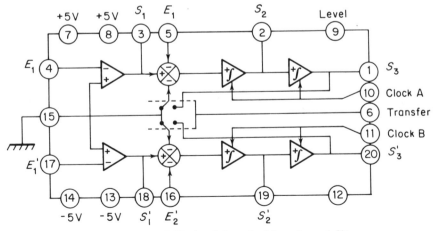

Figure 5.25 MF 10 circuit is a double universal filter

Table 5.2 Switched capacitor filter using MF10 device

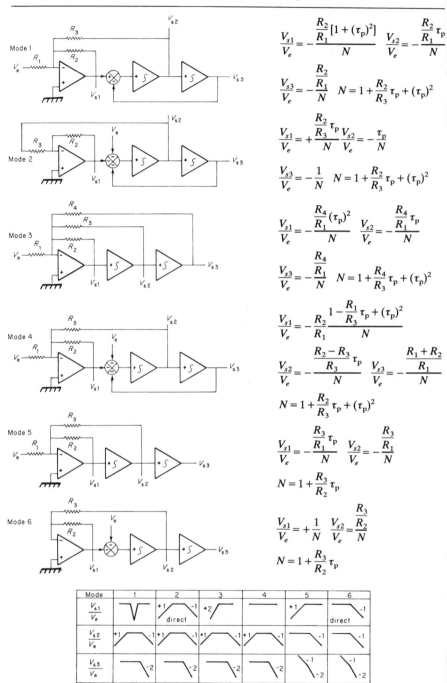

for the first time in 1978. Thus, it is obvious that the MF 10 retains some early design weakness, which should be improved in the future. In spite of this it displays major advantages compared to active filters:

— The filter time constant is an external variable which is the sampling frequency. This frequency can be controlled by the use of a VCO, for example.
— It is possible to achieve very low cut-off frequencies while keeping a very low level distortion.
— It is possible to realize very compact high-order filters. A single MF 10 gives a fourth-order filter when wiring both ways in series.
— Lower cost than other technologies (MF 10 costs about 40 US dollars).
— The filter characteristic properties depend very little on external parts of the filter.

Table 5.2 indicates several examples of possible wiring of this filter.

5.2 ANALOGUE MULTIPLIERS

5.2.1 Definition—Generalities

An analogue multiplier is a module that produces an output signal proportional to the product of both input signals X and Y.

$$S = \frac{XY}{K}$$

K is called the *scale factor*. If fixed, it is commonly equal to 10 V, but it can be adjustable. If the input may be of either positive or negative polarity and the output is in the correct relationship for multiplication, the device is called a four-quadrant multiplier. If only one input may be either positive or negative, the device is called a two-quadrant multiplier.

Numerous technologies can be used to realize these devices; for example:

— the quarter-square method, using parabolic characteristics achieved by diode function generator circuits;
— magneto resistive effect
— Hall effect;
— modulation of triangle signals;
— pulse modulation;
— logarithmic conversion;
— bipolar transistor transconductance.

The last two processes give the best performance and are commonly used in monolithic devices. The transconductance multiplier has a wide bandwidth (up to 200 MHz) while a good accuracy (maximal error with a full-scale input 0.1%) is realized with logarithmic multipliers. Multipliers

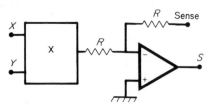
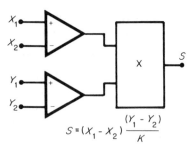

Figure 5.26 The sense terminal enables to operate a voltage–current conversion on a setting of output level

Figure 5.27 Differential input amplifier

$$S = (X_1 - X_2)\frac{(Y_1 - Y_2)}{K}$$

play a major role in analogue processing of measured signals. They are used for modulation and demodulation, fixed and variable remote gain control, power measurement, and mathematical operation in analogue computing, curve-fitting, and linearizing.

In most cases, multiplying blocks include operational amplifiers. It permits the realization of independent blocks performing various processes. (Figures 5.26–28 show the main structures.)

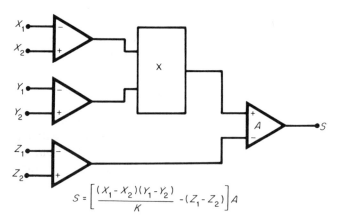

$$S = \left[\frac{(X_1 - X_2)(Y_1 - Y_2)}{K} - (Z_1 - Z_2)\right]A$$

Figure 5.28 General-purpose multiplier

Some manufacturers produce a multiplier and a divider integrated in a single block. These modules are called *multifunction devices*. The output voltage of such modules is connected to the three input voltages by the operation:

$$S = KY\left(\frac{Z}{X}\right)^m$$

where S may be either positive or negative and K is the scale factor. The power m can be adjusted between 0.2 and 5.

This module, which uses logarithmic converter principles, includes a block giving the logarithm of Y, a part giving the logarithm of Z/X, a summing block realizing the following operation:

$$K\left[m \log \frac{Z}{X} + \log Y\right],$$

and finally an exponential converter to obtain the final result.

A large number of functions are available from this module. As well as multiplication and division, it is possible to obtain the exponential function, squaring, square root extraction, sine, cosine, arctan, vector modulus, etc.

Example of industrial device The inner functional structure of the 4301 multifunction device from Burr Brown is shown in Figure 5.29.

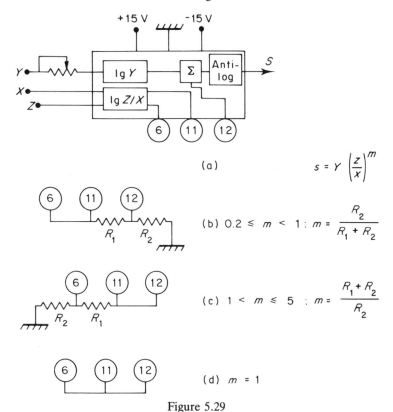

Figure 5.29

5.2.2 Using Multipliers in Signal Processing

(a) *Squaring.* For this operation $X = Y$. This leads to one of the most important uses in signal processing: the calculation of a signal r.m.s. value.

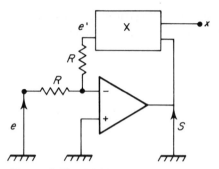

Figure 5.30 Division of two voltages

(b) *Division.* The particular wiring of the output amplifier described in Fig. 5.30 enables division of two voltages:

$$S = -K\frac{e}{X} \quad \text{because} \quad e = -e' \quad \text{and} \quad e' = \frac{XS}{K}.$$

It is important to notice that X must be kept positive in order to preserve stability. This circuit can be used as a voltage-controlled gain amplifier, with gain varying from 1 to ∞.

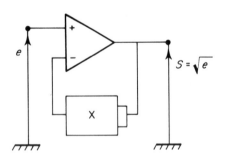

Figure 5.31 Square root

(c) *Square root operation.* The output is equal to the square root of the input. The input e must be positive, otherwise the circuit is unstable.

(d) *Linearization.* The calibration curve of a sensor is not necessarily linear. Furthermore, a sensor is commonly employed in the nonlinear part of its characteristic curve in order to improve the measuring sensitivity. This procedure implies a further linearization. Let us go back to the Wheatstone bridge circuit discussed in Chapter 2. The sensor is one of the bridge

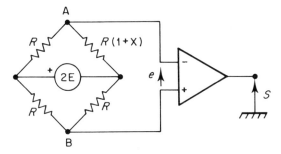

Figure 5.32

resistors. When its value increases from R to $R(1+x)$, the amplitude of the measured physical phenomenon is x. We can write

$$V_A = 2E\frac{1+x}{2+x}$$

$$V_B = E$$

so that

$$e = V_A - V_B = E\frac{x/2}{1+(x/2)}.$$

When the bridge is unbalanced the voltage, which is produced, is equal to

$$e = E\frac{x/2}{1+(x/2)} = E\frac{X}{1+X} \quad \text{with} \quad X = \frac{x}{2}$$

As the relationship between X and e is not linear, linearization is needed. It can be realized by using the circuit of Figure 5.33. For a simpler case, we assume that $E = K$.

(e) *Voltage-programmable filters.* When including a multiplier on the input of a circuit using an inverting operational amplifier (Figure 5.34), if e is the input sine wave and V_c the d.c. control voltage, the multiplier output is equal to

$$E = \frac{eV_c}{K} \quad \text{and so} \quad i = e\frac{V_c}{KR}.$$

It is as if e were applied to a voltage-controlled resistor. Its value follows

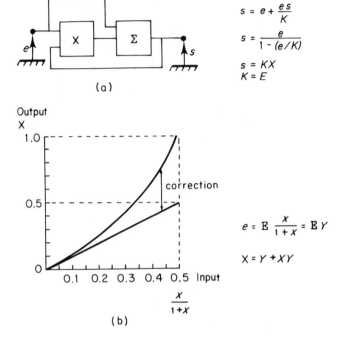

X	$Y = X/(1+X)$	Correction
0	0	0
0.0101	0.01	0.0001
0.0204	0.02	0.0004
0.0526	0.05	0.0026
0.1111	0.10	0.0111
0.25	0.20	0.05
0.429	0.30	0.129
0.667	0.40	0.267
1.00	0.50	0.500

Figure 5.33

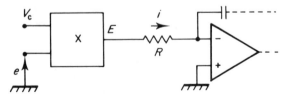

Figure 5.34 It is as if e were applied to a V_C voltage-controlled resistor

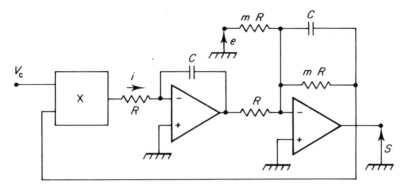

Figure 5.35 The filter central frequency is voltage controlled

the equation

$$R' = \frac{KR}{V_c}$$

Figure 5.35 indicates an example of a band-pass filter where the central frequency is voltage controlled by V_c.

(f) *Amplitude modulation.* A multiplier enables easy modulation of any kind of signal

$$v(t) = Ax(t)\cos(\omega_0 t + \phi_0),$$

where $x(t)$ is the modulating signal and $A\cos(\omega_0 t + \phi_0)$ is the carrier signal.

In that case we realize suppressed carrier modulation. Including a summing block at the useful signal input gives amplitude modulation with carrier as shown on Figure 5.36. Both modulation systems are depicted in Figure 5.37. In case of modulation with the carrier unsuppressed, the useful signal is obtained using the process which separates the two envelopes. In the case of suppressed carrier modulation it is not possible to restore $x(t)$ from one of these envelopes.

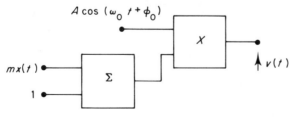

$v(t) = A[1 + mx(t)]\cos(\omega_0 t + \phi_0)$

Figure 5.36 Amplitude modulation with carrier

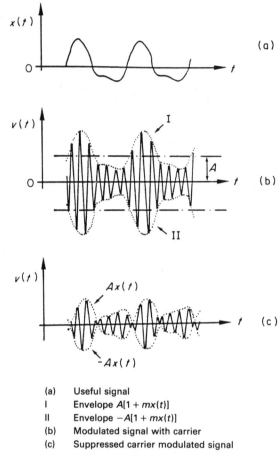

(a) Useful signal
I Envelope A[1 + mx(t)]
II Envelope −A[1 + mx(t)]
(b) Modulated signal with carrier
(c) Suppressed carrier modulated signal

Figure 5.37 Both amplitude modulation types

(g) *Demodulation.* The signal to be demodulated is in most cases a low-frequency signal:

$$x(t) = A \sin 2\pi Ft.$$

Let us assume the following expression to describe the modulated signal:

$$v(t) = A \sin 2\pi Ft \sin 2\pi ft.$$

In order to restore $x(t)$ from $v(t)$ the following product is calculated

$$v(t) \sin 2\pi ft = A \sin 2\pi Ft \sin^2 2\pi ft$$
$$= A \sin 2\pi Ft \left[\frac{1}{2} - \frac{\cos 4\pi ft}{2}\right].$$

But $4\pi f$ is assumed to be much greater than $2\pi F$ and simple low-pass filtering permits restoration of the useful signal $(A/2)\sin 2\pi Ft$.

This process is called *synchronous demodulation*. It is commonly used in various forms.

Application example

Assume that a phase difference measuring system has to be designed and the characteristics of the signal are described as follows:

— amplitude and signal frequency varying in a large range.
— signals being lost in noise.

The synchronous detection needs a delay line as time standard. It operates as follows:
The measurement system produces the following signals:

$$e_1(t-\tau) = E_1 \cos \omega(t-\tau)$$

$$p(t,\tau) = E \cos(\omega t + \phi) \cos \omega(t-\tau) \quad \text{with} \quad E = E_1 E_2$$

$$m(\tau) = \frac{1}{T}\int_0^T p(t,\tau)\,dt \quad \text{with} \quad T = \frac{2\pi}{\omega}.$$

This gives

$$m(\tau) = \tfrac{1}{2}E \cos(\phi + \omega\tau). \tag{5.3}$$

τ is trimmed until $m(\tau_0) = 0$, which enables the determination of ϕ from

$$\phi + \omega\tau_0 = \varepsilon(2K+1)\frac{\pi}{2} \tag{5.4}$$

with $\varepsilon = \pm 1$ and K an integer ≥ 0. Figure 5.38 describes the whole measurement system.

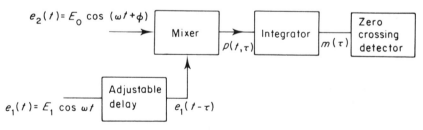

Figure 5.38 Phase measurement system using synchronous detection

The determination of ϕ implies the measurement of $\omega\tau_0$. But remaining doubt concerning ε and K will disappear with knowledge of the sign of $dm(\tau)/d\tau$ (Figure 5.39). For example, let us assume $-\pi < \phi < \pi$. It is possible to obtain positive or negative value of τ when it is noted that both input signals are inverted.

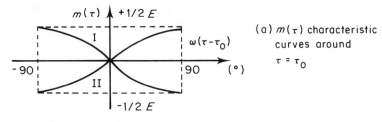

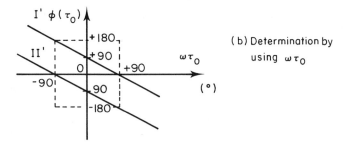

Figure 5.39 Curves for phase determination

If $K = 0$ the problem is simpler and enables plotting of the characteristic curves of Figure 5.39.

— Figure 5.39(a) are plotted from expression (5.3). Around $\tau \simeq \tau_0$ they are symmetrical. The sign of $dm(\tau)/d\tau$ enables us to choose the curve corresponding to the measurement.
— Figure 5.39(b) are plotted from expression (5.4). Knowing the value of $\omega\tau_0$, ϕ is calculated using:

· curve I' for negative values of $dm(\tau)/d\tau$
· curve II' for positive values of $dm(\tau)/d\tau$.

— The measurement of ϕ is then made without any ambiguity.

Such a system has numerous advantages:

(a) Ease of operation: the mixer may be an analogue multiplier; the integrator may be a simple RC circuit.
(b) The resulting measure doesn't depend on input signal level (see expression (5.3) and (5.4)).
(c) Good sensitivity: Assume that $\phi' = w(\tau - \tau_0)$. The curves of Figure

5.39 then give

$$\left|\frac{dm(\tau)}{d\phi'}\right| = \tfrac{1}{2}E \cos \phi'$$

the peak slope is obtained for $\phi' = 0$. We have in that case:

$$\left|\frac{dm(\tau)}{d\phi'}\right|_{\phi'=0} = \tfrac{1}{2}E$$

Numerical example With an input level $E = 100$ mV the sensitivity is then 50 mV/rad or 0.87 mV/degree and a standard nanovoltmeter detects $10\,\mu$V. It enables the detection of a 0.01 degree phase variation.

(d) *Noise immunity.* Suppose that a white noise $n(t)$ is added to $e_2(t)$ and that cross-correlation between $n(t)$ and $e_2(t)$ is zero, meaning that the noise $n(t)$ is statistically independent of $e_2(t)$.

Let us examine the following noise $n'(t) = n(t) \cos \omega(t - \tau)$ at the integrator input. The autocorrelation function of such a noise is given by

$$C_{n'n'}(\theta) = \tfrac{1}{2}C_{nn}(\theta) \cos \omega\theta = \tfrac{1}{2}N_F^2 \delta(\theta) \cos \omega\theta$$

with

$C_{nn}(\theta)$ = autocorrelation function of $n(t)$

N_F^2 = power spectral density of $n(t)$

$\delta(\theta)$ = Dirac's impulse.

Fourier transform of $C_{n'n'}(\theta)$ then gives the power spectral density $N_F'^2$ of $n'(t)$ as

$$N_F'^2 = \tfrac{1}{2}N_F^2$$

We are able to deduce the noise at the integrator output, by taking the inverse Fourier transform of $N_F''^2$ to obtain

$$N_F''^2 = N_F'^2 \frac{f_c^2}{f_c^2 + f^2}$$

with

$$f_c = \frac{1}{2\pi RC} \qquad f = \frac{\omega}{2\pi}.$$

The autocorrelation function $C_{n''n''}(\theta)$ of $n''(t)$ is given by

$$C_{n''n''}(\theta) = \tfrac{1}{2}N_F^2 \pi f_c \exp(-2\pi f_0 |\theta|).$$

This infers then the r.m.s. noise voltage at integrator output is

$$n''_{\text{rms}} = N_F \sqrt{\left(\frac{\pi f_c}{2}\right)}$$

so that finally the signal/noise ratio R_S at the output is

$$R_S = \left|\frac{m(\tau)}{n''_{rms}}\right|_{\tau=\tau_0} = \frac{E|\phi'|}{N_F\sqrt{(2\pi f_c)}} \qquad (5.5)$$

Numerical example A 40 dB R_S is needed. Assume that $E = 100$ mV, $|\phi'| = 3 \times 10^{-4}$ radians and N_F thernal noise in a 1 MΩ resistor. We have:

$$N_F^2 = \tfrac{1}{2}kTR \quad \text{at } 20°\text{C},$$

with
$k = 1.38 \times 10^{-23}$ J/K   Boltzmann constant
R (Ω)   resistor value
T (K)   thermodynamic temperature

$$N_F^2 = \tfrac{1}{2}1.38 \times 10^{-23} \times 293 \times 10^6 = 20 \times 10^{-16} \text{ V}^2/\text{Hz}.$$

The R and C parts of the integrator should be chosen so that

$$f_c = \frac{E^2|\phi'|^2}{2\pi N_F^2 R_S^2} = 8 \text{ Hz}.$$

There are two conflicting requirements that will have an effect on the RC time constant.

— It should be the highest possible to improve the signal/noise ratio.
— It should be small in order to perform fast measurement.

Expression (5.5) and the purpose of the system enable us to choose the optimum solution.

5.2.3 Multiplier Precision

In most cases, the manufacturer's specifications furnish the total error. This is defined by comparing the output and full-scale input values for X and Y (for example ± 10 V) theoretical and measured values.

The linearity error is calculated with the maximum difference between theoretical and real value of output S when one input value is kept constant while the other varies the offset voltage being compensated.

Some modules are fitted with a potentiometer for offset compensation. Offset voltage is present on input and output:

$$S = \frac{1}{K}(X + X_d)(Y + Y_d) + S_d$$

or:

$$S = \frac{XY}{K} + \frac{X_dY}{K} + \frac{Y_dX}{K} + \frac{X_dY_d}{K} + S_d$$

The terms $(X_d Y_d/K) + S_d$ create a constant offset. While both terms $X_d Y/K$ and $Y_d X/K$ show that even if one input is kept at zero, the output may still be proportional to the other input.

Notice that for analogue logarithmic multipliers the precision remains correct with low level inputs ($E < 100 \text{ mV}$).

5.3 LOGARITHMIC CONVERTERS

5.3.1 Definition

These modules provide an output voltage proportional to the logarithm of an input quantity. The ideal output equation is:

$$S = K \log \frac{E}{E_0},$$

where E_0 is a constant with the dimension of E.
K is the output scale factor having the dimension of S.
This function is commonly used in:
(a) Data compression.

Example When measuring the following mechanical impedance:

$$\bar{Z}_m = \frac{F \exp j\omega t}{V \exp [j(\omega t - \phi)]}$$

$F \exp j\omega t$ is the force applied at the observed structure, and $V \exp [j(\omega t - \phi)]$ is the velocity of the point where the force is applied.
The $\bar{Z}_m$ module may display considerable variations (sometimes 120 dB) between resonance and antiresonance. In this case log-converters are a classical solution.

(b) Plotting of a system transmittance module in Bode chart.
(c) Detection of very low currents.

Example The detected current at the output of photomultipliers is about a picoampere. It is often necessary to proceed to a log-conversion of input information in order to get a proper reading of this current.

(d) Linearizing the transducer signal. When the output is a logarithmic function of the measured variable the logarithmic device is connected for exponential operation. Whereas if the output is an exponential voltage, the device is connected for logarithmic operation.
(e) Improving signal/noise ratio in a transmission line.

Example The signal/noise ratio at the receiving station is

$$\left| \frac{E}{V_b} \right|$$

where E is the transmitted signal
V_b is the noise voltage.
If $E = 1 \text{ mV}$ and $V_b = -1 \text{ mV}$

$$\left| \frac{E}{V_b} \right| = 1$$

If a logarithmic conversion of the signal is performed before transmitting, the transmitted voltage will be:

$$V = K \log \frac{E}{E_0}.$$

If $E_0 = 100$ mV and $K = -2$ V, then we will have $V = 4$ V. At the receiving station we obtain $V + V_b = 3.999$ V. The antilogarithmic conversion, which is the reciprocal of the logarithmic conversion, restores a voltage with an absolute error of $1.1\,\mu$V. The signal/noise ratio is now about 870.

(f) A very interesting property of the logarithmic converter concerns the input error and output of the module. A fixed relative error will produce a fixed absolute error at the output.

Example Suppose the transfer equation of a logarithmic converter is:

$$S = 1\,\text{V} \times \log \frac{E}{E_0} \quad \text{(scale factor } K = 1\,\text{V)}$$

If a 1% variation of the input signal E is produced, the logarithmic converter output voltage can be written:

$$S = 1\,\text{V} \times \log \frac{E \times 1.01}{E_0}$$

or:

$$S = \underbrace{1\,\text{V} \times \log \frac{E}{E_0}}_{\text{initial value}} \pm \underbrace{1\,\text{V} \times \log 1.01}_{\substack{\text{absolute error due} \\ \text{to input variation}}}$$

this absolute error for a 1% variation is constant and equal to ± 4.3 mV.

5.3.2 Industrial Design

The method approximating a theoretical function with straight line segments has been used for a long time but is now replaced in today's monolithic design. The realization of bipolar transistors which have rigorously logarithmic characteristic curve for the emitter base junction tends to generalize the following structure where it is associated with an operational amplifier.

Note The collector current of a bipolar transistor is given by

$$i_c = \alpha i_E = \alpha i_{ES} \left[\exp \frac{q}{kT} V_{BE} - 1 \right]$$

assuming that $\alpha = 1$ and i_{ES} very low ($\sim 10^{-13}$ A) this is well approximated by

$$i_c \approx i_{ES} \exp \frac{q}{kT} V_{BE}.$$

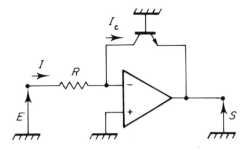

Figure 5.40 Basic log converter circuit

In the basic circuit of Figure 5.40 the output voltage is given by:

$$s = -\frac{kT}{q}\ln\frac{E}{Ri_{ES}},$$

where
 i_{ES} = reverse current of the saturated emitter-to-base junction
 k = Boltzmann constant, 1.38×10^{-23} JK^{-1}
 T = absolute temperature of the junction, K
 q = electron charge, 1.6×10^{-19} C.

What are the main problems appearing in practical realization?

The quality of the operational amplifier is decisive when converting very low level signals. In this following, we will assume the operational amplifier to be ideal.

5.3.2.1 Stability

Although the log converter module is not a linear system, it is possible to add, at fixed value of the input quantity, a low-level alternating signal for which the transmittance can be easily linearized. For this reason industrial manufacturers specify bandwidth and stability the way they are usually defined for linear systems.

Without becoming involved in theoretical details, let us point out that the frequency compensation allowed by the capacitor C and the resistance R_E in Figure 5.41 is necessary to preserve stability. Another point is the decreasing bandwidth for low-level variations with the average level of the input signal. It decreases when the average input current decreases.

5.3.2.2 Temperature compensation

The precision of the logarithmic converter is strongly dependent on temperature due to both terms kT/q which varies proportionally to T and i_{ES} which doubles at every 8°C change.

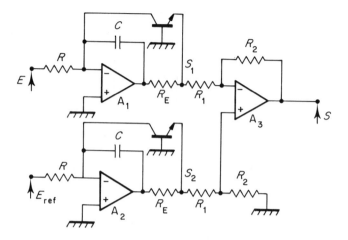

Figure 5.41 Elimination of I_{ES} current influence

(1) Compensation of i_{ES} Variations

The simple solution which is illustrated in Figure 5.41, uses two identical converters. The first one has an input voltage to convert (E), whilst the second one has a reference voltage (E_{ref}). The output voltages are:

$$S_1 = -\frac{kT}{q}\ln\frac{E}{Ri_{ES1}}$$

and

$$S_2 = -\frac{kT}{q}\ln\frac{E_{ref}}{Ri_{ES2}}.$$

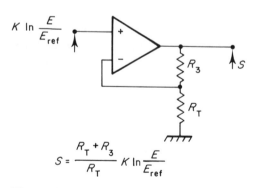

Figure 5.42 Compensation of scale factor variations

The final operational amplifier A3 provides

$$S = \frac{R_2}{R_1}\frac{kT}{q}\ln\frac{E}{E_{\text{ref}}}.$$

The term $-\ln(I_{\text{ES1}}/I_{\text{ES2}})$ is eliminated if both transistors are included in an isothermic block. Notice that the scale factor K

$$K = \frac{R_2}{R_1}\frac{kT}{q}$$

is linked with the basis of the logarithm.

(2) Compensation of Scale Factor Variations $K = (R_2/R_1)(kT/q)$
This factor creates considerable drift of the output voltage (0.33% K^{-1} at 27°C). The compensating circuit uses a resistor directly proportional to temperature which, once associated with R_T in Figure 5.43 generates an opposite drift (-0.33% K^{-1} at 27°C).

These two temperature-compensating systems are fabricated in today's devices according to the circuit of Figure 5.43, which requires only two operational amplifiers.

The different parameters are linked by

$$E_1 = -\frac{kT}{q}\ln\frac{E}{RI_{\text{ES1}}}$$

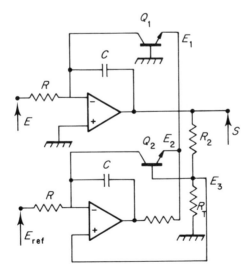

Q_1, Q_2 transistors

Figure 5.43 Log converter

and

$$E_3 - E_2 = E_3 - E_1 = \frac{kT}{q} \ln \frac{E_{ref}}{Ri_{ES2}}.$$

Thus

$$E_3 = \frac{kT}{q} \ln \frac{E_{ref}}{Ri_{ES2}} - \frac{kT}{q} \ln \frac{E}{Ri_{ES1}}$$

$$= -\frac{kT}{q} \ln \frac{E}{E_{ref}} \frac{i_{ES2}}{i_{ES1}}$$

and then

$$S = -\frac{R_T + R_2}{R_T} \frac{kT}{q} \ln \frac{E}{E_{ref}} \quad \text{with} \quad \frac{i_{ES1}}{i_{ES2}} = 1.$$

Note This circuit will calculate the logarithm of a ratio if E_{ref} is replaced by another input signal.

It allows realization of the reverse conversion, i.e. an antilogarithmic or exponential generation, by a simple rearrangement of the circuitry. An exponential output from a linear input is provided by circuit of Figure 5.44:

$$S = \frac{R_1}{R} E_{ref} \exp\left(-\frac{E}{K'}\right) \quad \text{with} \quad K' = \frac{kT}{q}\left(1 + \frac{R_2}{R_T}\right).$$

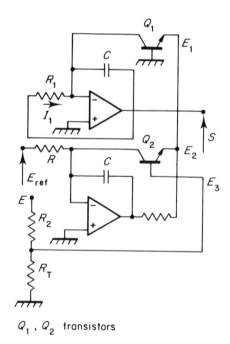

Q_1, Q_2 transistors

Figure 5.44 Antilog converter

In effect we can write

$$E_3 = E \frac{R_T}{R_T + R_2}$$

$$E_1 = E_3 - \frac{kT}{q} \ln \frac{E_{ref}}{R i_{ES2}} = -\frac{kT}{q} \ln \frac{i_1}{I_{ES1}}$$

$$S = R_1 i_1.$$

Therefore

$$S = R_1 \frac{E_{ref}}{R} \exp\left(-\frac{q}{kT} \frac{R_T}{R_T + R_2} E\right).$$

Beside the usual errors due to operational amplifiers (offset bias current, offset voltage, bandwidth, ...) the log converter is characterized by its correspondence with the logarithmic law, i.e. the difference between the output voltage and the logarithm of the input quantity. It is called the *log-conformity error*.

5.3.2.3 Example of monolithic package

Transfer characteristics

$$S = -K \log_{10} \frac{i}{I_{ref}}$$

Model 757 N and 757 P from Analog Devices.

— 6 decades from 1 nA to 1 mA.
— log-conformity error:

    0.5% between 10 nA and 100 μA
    1% between   1 nA and   1 mA

— Price per unit: 85 US dollars (1986).

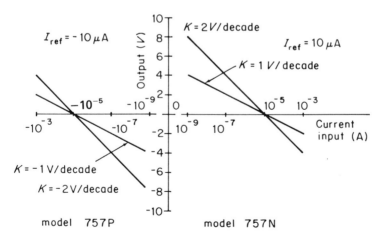

Figure 5.45

Antilog-Converter
Transfer characteristics

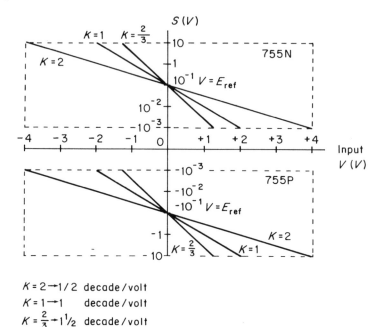

$K = 2 \rightarrow 1/2$ decade/volt
$K = 1 \rightarrow 1$ decade/volt
$K = \frac{2}{3} \rightarrow 1\frac{1}{2}$ decade/volt

Figure 5.46

$$S = E_{\text{ref}} \times 10^{V_{IN}/-K}$$

Model 755 N and 755 P from Analog Devices

— Log or antilog mode
— Current input

6 decades from 1 nA to 1 mA

— Voltage input

4 decades from 1 mV to 10 V

— 1% error max from 1 mV to 10 V
— Price per unit: 70 US dollars

5.4 A.C./D.C. CONVERTERS

Measuring an alternating current or voltage generally implies its conversion into its mean absolute value, its root mean square (r.m.s.) value or its peak-to-peak value.

5.4.1 Absolute-value Converters

The usual solution is a full-wave rectifier realized by a diode bridge. However, the forward voltage of the diodes ($V_d = 0.6$ V) prevents accurate conversion often needed in instrumentation. This effect can be avoided if the bridge rectifier is operated with a controlled current source as shown in Figure 5.47. The op amp is used as a voltage-controlled current source. Hence, current i_A flowing through the floating charge R_c follows the absolute value of the input voltage:

$$i_A = \frac{|e|}{R}.$$

It is independent of the diode forward voltage. This circuit is commonly used in analogue voltmeters; R_c is therefore replaced by a moving-coil ammeter. If ground-referenced output voltage is needed, it is possible to use a simple full-wave rectifier shown in Figure 5.48.

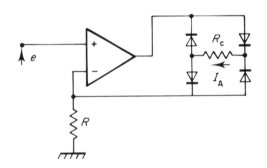

Figure 5.47 Full-wave rectifier

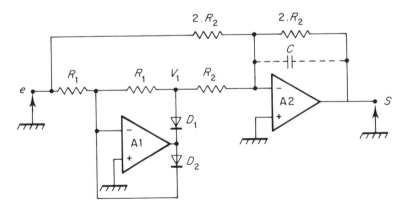

Figure 5.48 Full-wave rectifier with single-ended output

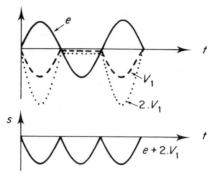

Figure 5.49 Operation of full-wave rectifier

Amplifier A1 operates as an inverting half-wave rectifier:

$$V_1 = \begin{cases} -e & \text{for} \quad e \geq 0 \\ 0 & \text{for} \quad e \leq 0 \end{cases} \tag{5.6}$$

whilst amplifier A2 computes the expression

$$S = -(e + 2V_1) \tag{5.7}$$

With Eq. (5.6), we can write:

$$S = \begin{cases} e & \text{for} \quad e \geq 0 \\ -e & \text{for} \quad e \leq 0 \end{cases}$$

The operation is illustrated by Figure 5.49.

Adding capacitor C, converts amplifier A2 into a first-order low-pass active filter so that the output voltage will then be

$$S = |\bar{e}|.$$

5.4.2 R.M.S. Converter

The root mean square value of a signal $e(t)$ is defined by:

$$V_{\text{rms}} = \sqrt{\left(\frac{1}{T}\int_0^T e^2(t)\,\mathrm{d}t\right)}, \tag{5.8}$$

where T is the measuring interval, which must be large compared to the longest period contained in the signal spectrum. Commonly used techniques are thermal effect conversion and analogue computing conversion.

5.4.2.1 Thermal Effect Conversion

The r.m.s. value of a voltage is equal to the direct voltage that would produce the same amount of heat in the same resistor.

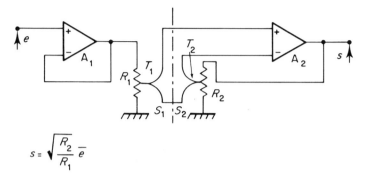

Figure 5.50 Thermal-effect conversion

S_1 and S_2 are both temperature sensors which measure the temperature of resistors R_1 and R_2. The input voltage is used to heat the resistor R_1.

Amplifier A2 keeps the temperature T_2 of R_2 equal to the temperature of R_1. If the values of thermal coupling between resistors in R_1 and R_2 are equal, we can write

$$\frac{s^2}{R_2} = \frac{\bar{e}^2}{R_1}, \quad \text{so} \quad s = \bar{e} \quad \text{for} \quad R_1 = R_2.$$

If thermocouples are used as thermal sensors, their low sensitivity requires a very accurate amplifier (low bias current, low drift), so a chopper amplifier is commonly used in such cases. The amplifier in this class can easily maintain drifts less than $0.1\,\mu\text{V}/^\circ\text{C}$.

If the temperature sensors are bipolar transistors, a less accurate amplifier may be needed for A2. In this case, the rise in temperature of the resistor R_1 will be detected by means of a change in V_{BE} of a transistor ($2\,\text{mV}/^\circ\text{C}$).

Example of industrial product The r.m.s. converter 4130 from Burr Brown is based on the above principle with Bipolar transistors used as temperature sensors. Conversion accuracy:

0.05% up to 100 kHz
2% up to 10 MHz.

5.4.2.2 Analogue computing converter

The r.m.s. value of an input signal $e(t)$ is obtained by realizing the analogue computation of Eq. (5.8). The operation of the converter, which is illustrated by Figure 5.51, consists of squaring the input voltage, taking the average and obtaining the square root. The conversion accuracy implies a high-level input voltage. If, for instance, the input voltage is equal to 1 mV and the scale factor K is 10 V as usual, a voltage of only $0.1\,\mu\text{V}$ is obtained

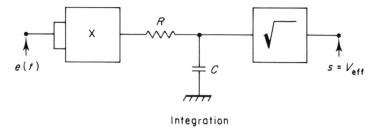

Figure 5.51 Analog computing conversion

at the output of the multiplier. Most of the time, this voltage is lower than the offset voltage of the squaring device and is drowned by the noise of the square rooter.

Example of industrial product
   Model 4340 from Burr Brown
   Conversion accuracy: ±2 mV ±0.2% reading.
   for input voltage 10 mV r.m.s. to 7.0 V r.m.s.
   100 Hz to 10 kHz sine wave.
It is possible to improve the conversion accuracy by the use of an external adjustment of gain and voltage offset.

Analogue computing conversion can also be performed using the logarithmic characteristic curve of bipolar transistors. The principle of operation of this kind of r.m.s. converter is illustrated by Figure 5.52. The input circuit provides a current i_1 proportional to rectified input voltage. The output voltage of amplifier A2 is expressed by

$$V = -2\frac{kT}{q} \ln \frac{|e|}{RI_s}.$$

Amplifier A4 operates as a logarithmic converter which provides $\ln s$ at its output. Then transistor T_2 produces a collector current i_2 proportional to

$$\ln^{-1}(\ln e^2 - \ln s) = \frac{e^2}{s}.$$

Finally amplifier A3 takes the average value of i_2 giving an output voltage which represents the r.m.s. value of $e(t)$.

Example of industrial product
   Model AD 536 from Analog Devices
   Conversion accuracy: 0.2% up to 20 kHz
                        1% up to 100 kHz
   Signal crest factor of 6 for 1% error
   Single power supply: from 5 to 36 V
   Approximate price per unit: 16 US dollars.

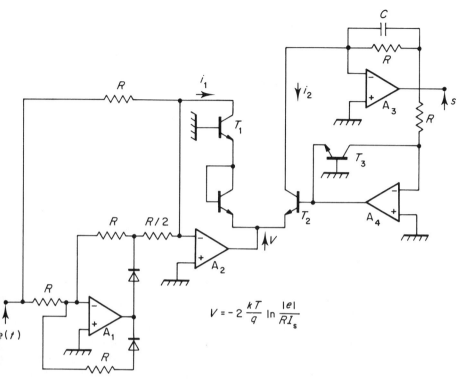

Figure 5.52   RMS converter using log characteristic of bipolar transistors

5.4.3 Peak-value Converter

The peak value of a voltage can be obtained by merely charging a capacitor through a diode; in order to suppress the error due to the offset voltage, the diode is included in the feedback loop of a follower according to Figure 5.53.

When $e(t) < V_c$, the diode isolates the capacitor from A1. When $e(t) > V_c$, we have $V_c = e(t)$ so that the capacitor is charged until the peak value is obtained. The follower A2 isolates the capacitor from the output whilst the analogue switch S is used to discharge the capacitor to prepare a new conversion cycle. R_1 suppresses oscillations produced by A1 charging the capacitor C.

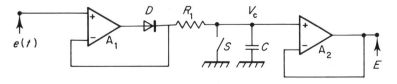

Figure 5.53   Peak value converter

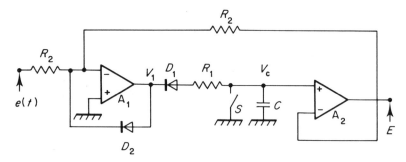

Figure 5.54 Improved peak value converter

The main drawback of this circuit is the saturation of A1 when $e(t) < V_c$ because the loop is then open. The recovery time due to this phenomenon limits the bandwidth of this circuit.

Figure 5.54 proposes a solution in order to increase the bandwidth of the device. Amplifier A1 operates as an inverter. When $e(t)$ becomes lower than $-V_c$, V_1 becomes negative and D_1 is then conducting. The feedback through both amplifiers produces $E = -e$. The offset voltage of the diode and the offset error of amplifier A2 are suppressed. When $e(t)$ decreases, V_1 increases, tends to clamp D_1 and opens the feedback loop made with R_2 while forward biassing D_2 whose conduction prevents the saturation of amplifier A1. The peak value is inverted and stored on capacitor C. There are no losses through D_1 and the infinite input impedance of A2. Reverse connection of D_1 and D_2 enables the conversion of negative peak values of the signal.

If the switch S is replaced by a resistor, a permanent measurement of the peak value of the input signal is obtained at the output. The resistor value should be calculated in order to give a negligible capacitor discharge between two successive peak values. A major drawback displayed by this system is a very slowly transmitted diminution of the peak value.

Some special applications require a fast measurement in a time shorter than the signal period. The measuring principle will then be based upon a calculation of the peak value of a sine signal $e(t) = E \sin \omega t$:

$$E = \sqrt{(E^2 \sin^2 \omega t + E^2 \cos^2 \omega t)} \qquad (5.9)$$

A differentiating circuit provides $\cos \omega t$ from $\sin \omega t$ as differentiating

$$V_1 = -RC \frac{de(t)}{dt} = ERC\omega \cos \omega t$$

If the frequency is a constant, $RC\omega$ can be made equal to 1. If not, an integrating circuit which suppresses the frequency-dependent $RC\omega$ term must be connected as shown on Figure 5.55. These various operations are

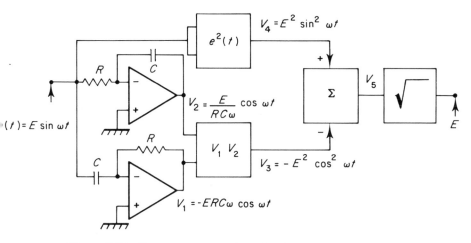

Figure 5.55 Permanent measuring of a sine wave peak value

illustrated by Figure 5.55, which includes modules already discussed in the preceding pages.

5.5 SYNCHRONOUS DEMODULATION

Synchronous demodulation or synchronous detection can be used to separate the useful sinusoidal signal from a noisy signal. It needs a controlled inverter as shown in Figure 5.56.

$$S = \begin{cases} e & \text{if} \quad V_c < V_p \\ -e & \text{if} \quad V_c = 0 \end{cases} \quad (V_p\text{: pinch-off voltage of the JFET})$$

The detected signal and the control signal have the same frequency and a constant difference in phase ϕ. The case with $\phi = 0$ is shown in Figure 5.57. V_c is a square signal. In that particular case the synchronous demodulation behaves like a full-wave rectifier.

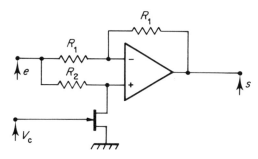

Figure 5.56 The JFET switch controls the output voltage sign

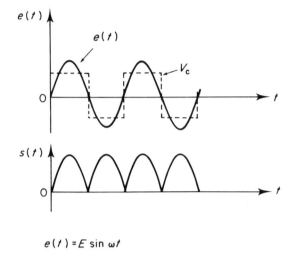

$e(t) = E \sin \omega t$

Figure 5.57 Synchronous demodulation with difference of phase ϕ equal to zero

In most cases with $\phi \neq 0$ the input signal $e(t)$ is multiplied by $+1$ or -1 according to the control signal at the frequency f_0. The output signal is described by the following equation:

$$s(t) = e(t)g(t)$$

with

$$g(t) = \begin{cases} 1 & \text{if} \quad V_c > 0 \\ -1 & \text{if} \quad V_c < 0 \end{cases}$$

The corresponding Fourier series is

$$g(t) = \frac{4}{\pi} \sum_{n=0}^{\infty} \frac{1}{2n+1} \sin(2n+1)\omega_c t$$

Let us suppose the input signal is a sine wave of frequency $f = mf_c$ and a phase difference ϕ with respect to the control signal. The output signal is then equal to

$$s(t) = E \sin(m\omega_c t + \phi) \frac{4}{\pi} \sum_{n=0}^{\infty} \frac{1}{2n+1} \sin(2n+1)\omega_c t$$

The average value of the output voltage is obtained by low-pass filtering and is equal to

$$\bar{S} = \begin{cases} \dfrac{2}{\pi m} E \cos \phi & \text{for} \quad m = 2n+1 \\ 0 & \text{for} \quad m \neq 2n+1. \end{cases}$$

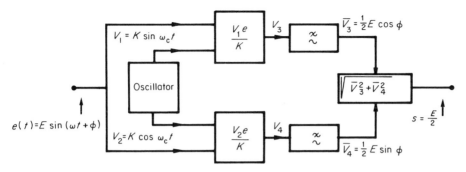

Figure 5.58 Synchronous demodulation independent of input signal phase $\left(S = \dfrac{E}{2} \text{ for } f = f_c\right)$

In the case of an input signal including harmonics only odd rank harmonics will contribute to the output value. This drawback can be suppressed by using an analogue multiplier instead of an analogue inverting switch and a sine wave $V_c \sin \omega_c t$ instead of square control signal V_c. Now a synchronous detection is performed.

If the amplitude of V_c is made equal to multiplier scale factor K, the output after appropriate filtering is

$$\bar{S} = \begin{cases} \dfrac{E}{2}\cos\phi & \text{if} \quad f = f_c \\ 0 & \text{if} \quad f \neq f_c \end{cases}$$

The output voltage can be made directly proportional to the input signal amplitude and independent of difference of phase ϕ in the circuit Figure 5.58. The oscillator in this circuit delivers two signals V_1 and V_2 with a phase difference of 90° and an amplitude equal to the scale factor K of both multipliers.

The output voltage is independent of angle ϕ. This circuit may be considered as a true selective voltmeter with a very high Q factor ($\sim 10^6$), If the oscillator frequency can be varied or tuned the circuit will behave like a spectrum analyser.

5.6 CORRELATORS

5.6.1 Principle

The autocorrelation product of a function $x(t)$ is the average value of the product of two values of $x(t)$ separated by a period of time τ. This product provides information concerning the correlation of values taken by $x(t)$ at

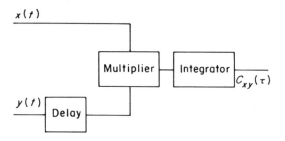

Figure 5.59 Analog correlator

instants separated by interval τ. A very powerful technology for narrow band signal detection is founded on this principle whose complexity has been removed due to the wide possibilities offered by modern electronics.

It is recalled that the correlation product between two functions $x(t)$ and $y(t)$ can be expressed by

$$C_{xy}(\tau) = \lim_{T \to \infty} \left\{ \frac{1}{T} \int_0^T x(t) y(t - \tau) \, dt \right\}.$$

This definition allows the establishment of the diagram of Figure 5.59 which looks like a synchronous detection circuit of Figure 5.38.

This simple procedure has two major drawbacks:

— It is very slow since the correlation function is obtained point by point.
— It is very difficult to realize important delay values with analogue components.

Most of today's correlators are digital which requires that both input signals be converted into sampled signals.

If the sampling period is T_s and the delay values are equal to a multiple of T_s, i.e. $\tau = mT_s$, the correlation function will then be equal to:

$$C_{xy}(mT_s) = \frac{1}{n} \sum_{k=0}^{k=n} x(kT_s) y[(k - m)T_s]. \qquad (5.10)$$

The computation is performed with the following steps:

— sampling and quantizing inputs $x(t)$ and $y(t)$;
— relative shifting of $y(t)$;
— product of samples;
— summing of products.

Two procedures are used in industrial devices.

5.6.2 Real-time Correlators

Both signals $x(t)$ and $y(t)$ are samples at the same time with the calculation progressing as shown in Figure 5.60.

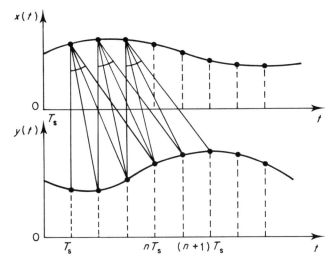

Figure 5.60 Real-time correlator

5.6.3 Row Correlators

The calculation is performed in the manner shown on Figure 5.61. Input $y(t)$ is sampled every T_s. Input $x(t)$ is sampled every $(n+1)T_s$. calculation has to be performed N times. Consequently, $(n+1)N$ samples are needed. The measuring time will then be $(n+1)NT_s$.

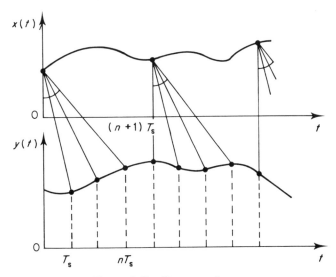

Figure 5.61 Row correlator

5.6.4 Industrial Devices

Figure 5.62 shows the simplified diagram of a correlator involving four different phases.

Phase 1. Both inputs are sampled, held and quantized, for example, on 9 different levels (-4 to $+4$). In fact, it may be proved that 9 different levels are sufficient for an accuracy higher than 10^{-4} on the correlation function. This property is due to the fact that this type of quantization does not change the characteristic function too much around the origin. Consequently it has little effect on, as the correlation function is its second derivative function at this point. The sampling period is T_s.

Phase 2. One of the inputs, $y(t)$, follows the delay channel. After quantization it is introduced into a 'measure block' which behaves like an N-stage shift register.

The delay period is T_s. Each new sampled value is stored in the first stage of the 'measuring sub-block', while the others are shifted by one stage. If

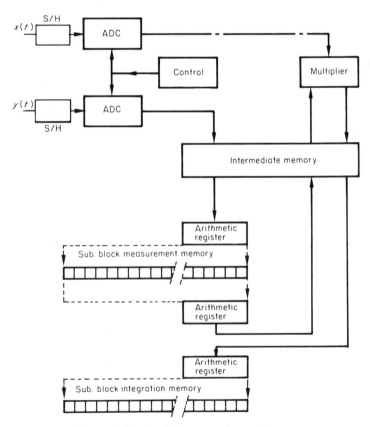

Figure 5.62 Real-time correlator diagram

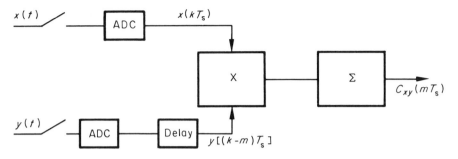

Figure 5.63 Digital correlator diagram

the number of sampled values exceeds N, those entered first are deleted successively.

Phase 3. When N sampled values of $y(t)$ are stored in the measure memory, during the delay the following products are simultaneously computed

$$\begin{array}{lll} X_N Y_N & \text{for} & \tau = 0 \\ X_N Y_{N-1} & \text{for} & \tau = T_s \\ X_N Y_{N-2} & \text{for} & \tau = 2T_s \\ \vdots & & \vdots \\ X_N Y_0 & \text{for} & \tau = NT_s. \end{array}$$

then X_N and Y_0 are deleted. Y_{N+1} is stored in the first stage of the sub-block and the same procedure is performed with the next sampled value X_{N+1}.

Phase 4. All the products possessing the same delay are summed in a digital integrator called the memory integration sub-block.

During the whole computation, an analogue switch cyclically sweeps all the summing memory stages. This allows the correlation function to be constructed.

This kind of correlator can operate with delay usually between a few microseconds and a few hundred seconds.

5.6.5 Example of application of correlation technique

Consider the following signal: $x(t) = s(t) + n(t)$, where $n(t)$ is noise and $s(t)$, the useful signal, is to be detected. It is assumed that knowledge of the function $C_{ss}(\tau)$ will be sufficient for this purpose, which is usually the case for periodic signals. The auto-correlation of $x(t)$ is obtained in the usual manner as

$$C_{xx}(\tau) = C_{ss}(\tau) + C_{nn}(\tau) + C_{ns}(\tau) + C_{sn}(\tau).$$

Assuming that noise and signal are uncorrelated, which practically means that $C_{ns}(\tau)$ and $C_{sn}(\tau)$ are negligible, $C_{xx}(\tau)$ becomes

$$C_{xx}(\tau) = C_{nn}(\tau) + C_{ss}(\tau).$$

As a noise autocorrelation function tends towards zero when τ increases them

$$\lim_{\tau \to \infty} C_{xx}(\tau) \to C_{ss}(\tau)$$

The signal/noise ratio, which changes with τ, can be represented by

$$\frac{C_{ss}(\tau)}{C_{nn}(\tau)}.$$

To obtain accurate detection, the signal must be periodic and the correlation time τ must be high so that the values of $C_{ss}(\tau)$ obtained give negligible loss of information on $s(t)$. If $s(t)$ is not periodic, its spectral bandwidth must be narrow compared to the noise bandwidth. This allows a reduction of the noise perturbation effect for very small values of τ. Practically, another cause of error is the finite integration time. This error contribution can be estimated in the case of a Gaussian noise distribution by using the following formula, which is valid for large τ values,

$$\left[\frac{S}{N}(\tau)\right]_{\text{output}} \approx \sqrt{(SNT)} \left[\frac{S}{N}\right]_{\text{input}}$$

where

T is the integration time expressed in seconds.
N is the noise bandwidth expressed in Hz.

When the shape of the signal to be detected is known, correlation may be realized by adaptive filtering of the input signal. This operation is achieved by cross correlating the input signal with a model. The level of likeness between these two signals can be measured by the correlator.

5.6.6 Monolithic Correlator

The correlation technique is becoming more accessible since the appearance of monolithic correlators. Let us mention an example: The TDC 1004 J type from TRW, which is a typical example, includes three 64-bit shift registers controlled by three separated clocks able to work up to 25 MHz. Both binary words shifted in registers A and B are continually compared bit by bit using an exclusive-OR gate network. The digital signals collected at the gate output are converted into analogue currents and summed to produce the output signal which represents the correlation between both binary signals applied through A_{in} and B_{in}. The third register (M) allows selection of the bit positions that are not taken into account by the correlation.

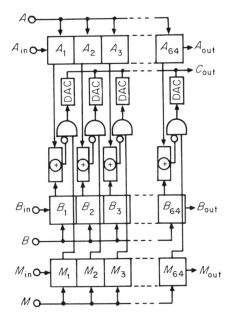

Figure 5.64 Monolithic correlator TDC 1004J from TRW

5.7 PHASE-LOCKED LOOP (PLL)

Although formerly used only by transmission systems, the phase-locked loop technology is now commonly in the measurement domain. Typical applications are as follows:

— demodulation of a frequency modulated signal,
— voltage-to-frequency conversion,
— frequency multiplication,
— tracking filters,
— frequency shifting,
— FSK coding/decoding (MODEM);
— synchronous demodulation of an amplitude-modulated signal,
— signal–noise separation;
— frequency synthesis.

5.7.1 Principle of Operation

Recall that the general expression of a wave is

$$s(t) = S \cos \phi(t).$$

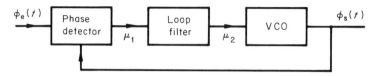

Figure 5.65 PLL schematic diagram

$\phi(t)$ is called the instantaneous phase of the signal

$$\omega(t) = \frac{d\phi(t)}{dt}$$

is called the instantaneous angular frequency of the signal. If the constant angular frequency ω_0, often called the *free running frequency*, is introduced then the signal may be expressed by

$$s(t) = S \cos[\omega_0 t + \theta(t)]$$

with

$$\phi(t) = \omega_0 t + \theta(t)$$

A phase-locked loop includes a phase comparator, a loop filter and a voltage-controlled oscillator (VCO).

5.7.1.1 The voltage-controlled oscillator

This oscillator produces a periodic signal whose frequency depends on voltage $\mu_2(t)$ at its input. If $\phi_s(t)$ is the instantaneous phase of the fundamental signal, the behaviour of the VCO is described by

$$\frac{d\phi_s(t)}{dt} = f[\mu_2(t)].$$

The derivative function

$$\left| \frac{d}{d\mu_2} f(\mu_2) \right| = k_\omega$$

expresses the *VCO sensitivity*.

5.7.1.2 The phase comparator

This produces a signal depending on the difference between instantaneous phases $\phi_e(t)$ and $\phi_s(t)$, and also depends on the shape and the amplitude. If $\phi_e(t)$ and $\phi_s(t)$ are both the phase of fundamental waves of these two signals the behaviour of the phase detector can be defined by

$$\mu_1(t) = g[\phi_e(t) - \phi_s(t)]$$

The derivative function

$$\left|\frac{d\mu_1(t)}{d[\phi_e(t) - \phi_s(t)]}\right| = k_\phi$$

expresses the *phase-comparator sensitivity*.

The function $g[\phi_e(t) - \phi_s(t)]$ is necessarily periodic because of the definition of the instantaneous frequency of a sine wave. Its period is 2π or a submultiple of 2π.

5.7.1.3 The loop filter

This is a low-pass filter. It is necessary to include it in the loop to insure stability of servo-control. The filter characteristics determine the loop performances in steady-state and transient operations.

The purpose of a phase-locked loop is to keep the phase ϕ_s of the oscillator output equal to the instantaneous phase of the input signal ϕ_e. Voltage μ_1 is a phase-error voltage when the loop is considered as in servo-theory. It controls the oscillator in order to keep ϕ_s equal to ϕ_e with a minimized phase error $\Delta\phi$.

5.7.2 Phase-locked Loop Operation For Sinusoidal Input

First of all, we assume that the phase-locked loop (PLL) is shown by its simplified diagram of Figure 5.66, in which the low-pass filter is missing. The omission of a filter does not affect the steady-state response of the loop.

Assume that the sensitivity of the phase detector and the voltage-controlled oscillator are constant. This hypothesis allows simplification of the operating equations.

The phase detector operation will be defined by

$$\mu_1(t) = k_\phi[\phi_e(t) - \phi_s(t)].$$

whilst the VCO operation functions in accordance with

$$\frac{d\phi_s(t)}{dt} = \omega_0 + k_\omega \mu_1(t),$$

where ω_0 is the *free-running frequency of the VCO*.

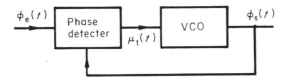

Figure 5.66 PLL without low pass filter

If the phase difference is equal to
$$\Delta\phi(t) = \phi_e(t) - \phi_s(t)$$
the loop equation is
$$\omega_0 + \frac{d[\phi_s(t)]}{dt} + k_\omega k_\phi \Delta\phi(t) = \frac{d\phi_e(t)}{dt}.$$
In the case of a sine wave, we have
$$\phi_e(t) = \omega_e t + \theta_e;$$
so that the expression for $\Delta\phi$ is then
$$\Delta\phi(t) = \frac{\omega_e - \omega_0}{k_\omega k_\phi} + C \exp(-k_\phi k_\omega t)$$
and the expression for $\phi_s(t)$ is
$$\phi_s(t) = \omega_e t + \frac{\omega_0 - \omega_e}{k_\omega k_\phi} - C \exp(-k_\phi k_\omega t).$$

The term $k_\omega k_\phi$ is the *loop gain*. The steady-state response of the loop to a sine wave of frequency ω_e is a sine wave having the same frequency ω_e but with a difference of phase
$$\Delta\phi = \frac{\omega_0 - \omega_e}{k_\omega k_\phi}.$$
which is proportional to the input angular frequency ω_e and decreases when the loop gain increases.

5.7.3 Practical Operation of a Phase-locked Loop

5.7.3.1 Choice of the phase detector

Suppose that $s_1(t)$ and $s_2(t)$, both sinusoidal with the same frequency, are given by
$$s_1(t) = a_1 \cos(\omega t + \theta_1)$$
$$s_2(t) = a_2 \cos(\omega t + \theta_2).$$
The product of these two sine waves is
$$s_1 s_2 = \frac{a_1 a_2}{2} \cos(\theta_1 - \theta_2) + \frac{a_1 a_2}{2} \cos(2\omega t + \theta_1 + \theta_2)$$
The average value of $s_1 s_2$ is then
$$\overline{s_1 s_2} = \frac{a_1 a_2}{2} \cos(\theta_1 - \theta_2)$$

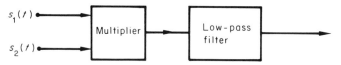

Figure 5.67 Analog phase detector

Thus phase detection may be achieved by using an analogue multiplier and a low-pass filter of cut-off frequency lower than ω. The characteristic curve of this detector is a sinusoid and its sensitivity is equal to $a_1 a_2/2$ when the phase difference is nearly equal to $(2p + 1)\pi/2$.

We notice that the product $a_1 a_2/2$ in

$$\frac{a_1 a_2}{2} \cos(\theta_1 - \theta_2)$$

is to be kept constant for correct phase detection. This implies constant amplitudes for both input signals.

Now, suppose that $s_1(t)$ is a sinusoid: $s_1(t) = a_1 \cos(\omega t + \theta_1)$ while $s_2(t)$ is a periodic signal of any shape. The Fourier series decomposition of $s_2(t)$ provides

$$s_2(t) = \overline{s_2(t)} + a_{21} \cos(\omega t + \theta_{21}) + \cdots + a_{2n} \cos(n\omega t + \theta_{2n} + \cdots$$

The product $s_1 s_2$ is then equal to

$$s_1 s_2 = a_1 \overline{s_2} \cos(\omega t + \theta_1) + \frac{a_1 a_{21}}{2} \cos(\theta_1 - \theta_{21})$$

$$+ \frac{a_1 a_{21}}{2} \cos(2\omega t + \theta_1 + \theta_{21}) + \cdots$$

$$+ \frac{a_1 a_{2n}}{2} \cos[(n - 1)\omega t + \theta_{2n} - \theta_1]$$

$$+ \frac{a_1 a_{2n}}{2} \cos[(n + 1)\omega t + \theta_{2n} + \theta_1] + \cdots.$$

But we still have the expression:

$$\overline{s_1 s_2} = \frac{a_1 a_{21}}{2} \cos(\theta_1 - \theta_{21}).$$

Notice that we can extract the phase difference between $s_1(t)$ and the fundamental component of $s_2(t)$ from the product $s_1 s_2$ and that the comparator characteristic curve is still a sinusoid. In the case when $s_2(t)$ is a square wave with amplitude a_2, the comparator sensitivity is equal to $a_1 a_2/\pi$ and the phase differences are nearly equal to $(2p + 1)\pi/2$.

Suppose now that both input signals are square signals with a zero average value. In such a case it is easy to calculate the average value of $s_1 s_2$.

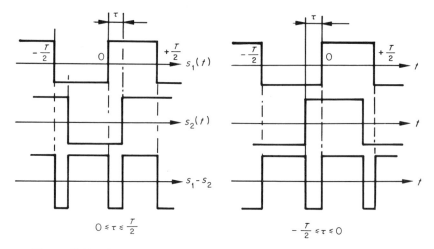

Figure 5.68 Operation diagram of a phase detector in case of two square wave input signals with zero average value

With the help of Figure 5.68, we demonstrate that the expression

$$\overline{s_1 s_2} = a_1 a_2 \left(1 - 4\frac{|\tau|}{T}\right) \quad \text{with} \quad \tau \in \left(-\frac{T}{2}; +\frac{T}{2}\right).$$

In terms of phase it gives

$$\overline{s_1 s_2} = a_1 a_2 \left(1 - 2\frac{|\Delta\phi|}{\pi}\right) \quad \text{with} \quad \Delta\phi \in (-\pi; +\pi).$$

This phase comparator displays a linear characteristic curve. As shown on Figure 5.69 it is linear only in an interval equal to π. The sensitivity is then constant in absolute value and is equal to:

$$\frac{2a_1 a_2}{\pi}.$$

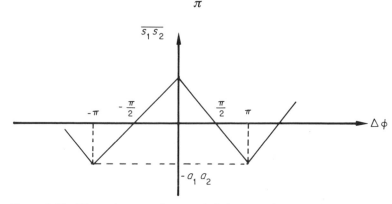

Figure 5.69 Phase detector characteristic in case of square input signals

Table 5.3

$s_1(t)$	$a_1 \cos(\omega t + \theta_1)$	$a_1 \cos(\omega t + \theta_1)$	$+a_1$ / $-a_1$ (square wave)
$s_2(t)$	$a_2 \cos(\omega t + \theta_2)$	$+a_2$ / $-a_2$ (square wave)	$+a_2$ / $-a_2$ (square wave)
Detector characteristic	Sinusoid	Sinusoid	Linear
Sensitivity	$\dfrac{a_1 a_2}{2}$	$\dfrac{a_1 a_2}{\pi}$	$2\dfrac{a_1 a_2}{\pi}$
Fundamental angular frequency	2ω	ω if $\overline{s_2} \neq 0$ else 2ω	2ω

According to the charts of Figure 5.68 the frequency of the fundamental component of the signal $s_1 s_2$ is equal to 2ω. This is no longer true if the duty cycle of one of the input square waves is not equal to 50%. For indeterminate duty cycles the phase detector depicted in Figure 5.67 cannot be employed.

Table 5.3 summarizes the phase detector properties considering wave forms on the multiplier input.

Beside the multiplying phase detector, it is interesting to mention three other commonly used detector circuits:

(1) Exclusive-OR gate phase detector
(2) Flip-flop phase detector
(3) Sample-and-hold phase detector

Sample-and-hold comparator operation can be explained as follows. A sawtooth voltage is started synchronously with the positive edge of one signal, then sampled synchronously with the positive edge of another signal. The voltage held in memory after a period T is directly proportional to the delay τ and thus to the phase difference between both signals.

The choice of phase comparator is essentially determined by the following considerations:

— operating frequency domain;
— input signal wave forms;

Analogue multiplier operation is limited to about 100 MHz for special İC devices called *balanced mixers*. Those circuits operate with both sine or square wave forms. At higher frequencies, sampling techniques can be used

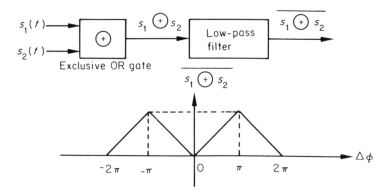

Figure 5.70

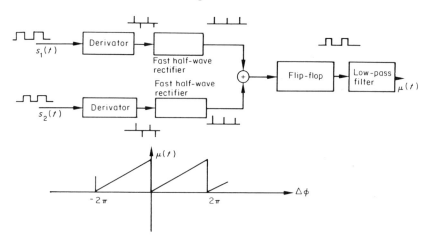

Figure 5.71 Flip-flop phase detector

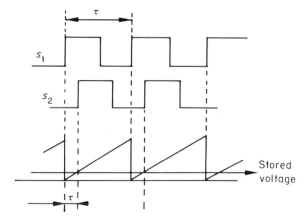

Figure 5.72 Sample-and-hold phase detector

up to a few GHz. For very high frequencies, sampling has to be performed by Schottky diode analogue switches.

The fastest digital circuits, but also the most expensive, are characterized by a propagation time below 1 ns. This fact allows the design of phase comparators operating up to some hundreds MHz; exclusive-OR gate detectors commonly require square input signals while sequential logic comparators (flip-flops) can operate with an input signal of any wave form.

5.7.3.2 Choice of the voltage-controlled oscillator

Voltage-controlled oscillators belong to the class of either multivibrators or sine wave generators. The first type displays a good linearity of modulation, a large wide band and produce also square, triangle or sine wave (with a few per cent distortion). They are easily trimmed because their free-running frequency is only defined with an RC circuit. Their main drawback is their frequency limitation to some tens of megahertz.

5.7.3.3 Choice of the loop filter

Table 5.4 gives the diagram and the transmittance of the three most commonly used filters.

Table 5.4

		Transmittance
First-order low-pass filter (I)	R, C circuit	$K(p) = \dfrac{1}{1 + \tau p}$ $\tau = RC$
Passive low-pass filter with feed-forward correction (II)	R_1, R_2, C circuit	$K(p) = \dfrac{1 + \tau_1 p}{1 + \tau_2 p}$ $\tau_1 = R_2 C$; $\tau_2 = (R_1 + R_2)C$
Active low-pass filter with feed-forward correction (III)	R_1, R_2, C, op-amp circuit	$K(p) = \dfrac{1 + \tau_1 p}{\tau_2 p}$ $\tau_1 = R_2 C$; $\tau_2 = R_1 C$

5.7.4 Loop Stability

Using the Laplace transform, we can write:

— for the phase detector: $U_1(p) = k_\phi[\phi_e(p) - \phi_s(p)]$
$$U_1(p) = k_\phi \Delta\phi$$

— for the filter: $K(p) = \dfrac{U_2(p)}{U_1(p)}$

— for the VCO: $\phi_s(p) = \dfrac{k_\omega}{p} U_2(p).$

We can draw the loop flow chart of Figure 5.73 to obtain the loop parameters.

The loop transmittance is

$$H(p) = \frac{k_\omega k_\phi K(p)}{p + k_\omega k_\phi K(p)}.$$

The open loop transmittancee is

$$B(p) = \frac{k_\omega k_\phi}{p} K(p).$$

The loop becomes unstable for any frequency satisfying the conditions

$$B(j\omega) = -1 \leftrightarrow \begin{cases} |B(j\omega)| = 1 \\ \arg B(j\omega) = \pi + 2k\pi. \end{cases}$$

The loop stability for the three filters in Table 5.4 may be examined using the gain characteristic curve plotted in Bode's chart form. By recalling the rule that a system including feedback is stable as long as the gain plotted in Bode's chart crosses the 0 dB axis with a slope of -1 (-20 dB per decade).

5.7.4.1 Filter 1: $K(p) = 1/(1 + \tau p)$

The value of τ is RC. The loop gain is then:

$$B(p) = k_\omega k_\phi \frac{1}{1 + \tau p}.$$

$B(j\omega)$ is plotted in Figure 5.74. The loop remains stable for $k_\omega k_\phi < \dfrac{1}{\tau}$.

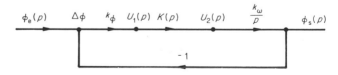

Figure 5.73 Loop flow chart

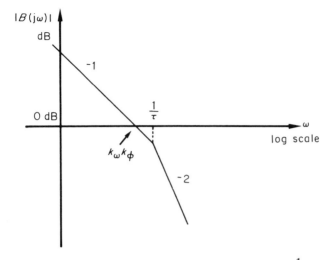

Figure 5.74 Asymptotic plotting of $B(j\omega) = k_\omega k_\phi \dfrac{1}{j\omega(1+\tau j\omega)}$

5.7.4.2 Filter 2: $K(p) = (1+\tau_1 p)/(1+\tau_2 p)$

The loop gain is then

$$B(p) = k_\omega k_\phi \frac{1+\tau_1 p}{p(1+\tau_2 p)}.$$

$B(j\omega)$ is plotted in Figure 5.75. The loop remains stable for

$$\frac{1}{\tau_1} < k_\omega k_\phi \frac{\tau_1}{\tau_2}$$

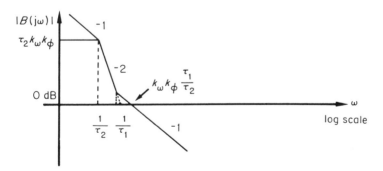

Figure 5.75 Asymptotic plotting of $B(j\omega) = k_\omega k_\phi \dfrac{1+\tau_1 j\omega}{j\omega(1+\tau_2 j\omega)}$

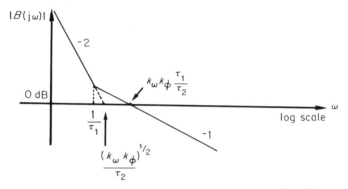

Figure 5.76 Asymptotic plotting of $B(j\omega) = k_\omega k_\phi \dfrac{1+\tau_1 j\omega}{\tau_2(j\omega)^2}$

5.7.4.3 Filter 3: $K(p) = (1 + \tau_1 p)/\tau_2 p$

The loop gain is then

$$B(p) = k_\omega k_\phi \frac{1+\tau_1 p}{\tau_2 p^2}$$

$B(j\omega)$ is plotted in Figure 5.76. The loop remains stable for

$$\frac{1}{\tau_1} < k_\omega k_\phi \frac{\tau_1}{\tau_2}.$$

We can infere that filters 2 and 3 allow a higher d.c. loop gain and consequently a higher loop velocity.

5.7.5 Locking Frequency

If a signal is applied to the input, and the angular frequency of the oscillator output ω_s is equal to the instantaneous angular frequency of the input signal ω_e, locking occurs. Locking frequencies are limited by the highest and lowest frequency of the VCO. With no signal input it oscillates freely at ω_0. When a frequency ω_e is suddenly applied to the input, a beat note is generated. The frequency of $\mu_1(t)$ is $\Delta\omega = \omega_e - \omega_s$. If $\Delta\omega$ is higher than the cut-off frequency of the filter, the VCO still oscillates at ω_0. But if $\Delta\omega$ is lower, the filter output signal shifts the oscillator frequency in order to maintain its instantaneous frequency $\phi_s(t)$ on the instantaneous frequency of the input signal $\phi_e(t)$.

5.7.6 Examples of Industrial Devices

PLL technique can be used in a very wide variety of applications in both analogue and digital systems. The low price of monolithic PLL makes them

Table 5.5

	LM 565 National semiconductor	XR 215 Teletek Airtronic
Maximal VCO frequency	500 kHz	35 MHz
VCO Sensitivity	6.6 kHz/V	adjustable
Phase-detector sensitivity	0.68 V/rad	2 V/rad

more and more common in the instrumentation field. Table 5.5 gives some examples.

5.7.7 Example of Operation

5.7.7.1 Digital measuring system for phase difference

Many digital phasemeters are available today. Although their steady-state performance characteristics are good they may not be sufficient for quick measurement, especially in the measurement phase of transient signals. Those available are analogue devices which operate an analogue-to-digital conversion and only display the result. It is obvious that several periods are necessary to perform an accurate measurement. Commonly they achieve the integration of a d.c. voltage during the whole dephasing, and take the average value computed for one period.

Figure 5.78 is a proposal for a fully digital phase-measuring system based upon the impulse counting principle with a PLL operating as a frequency multiplier. A frequency multiplier is released by the insertion of a frequency divider in the loop between the VCO and the phase detector. The loop acts to maintain the input frequency equal to the output frequency divided by n. When the loop is locked, we have $\omega = \omega_1$, but as

$$\omega_1 = \frac{\omega_s}{n}, \quad \text{so} \quad \omega_s = n\omega.$$

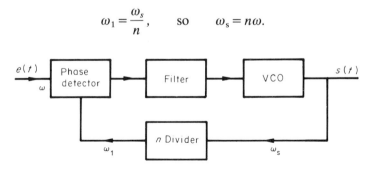

Figure 5.77 Frequency multiplier using a PLL

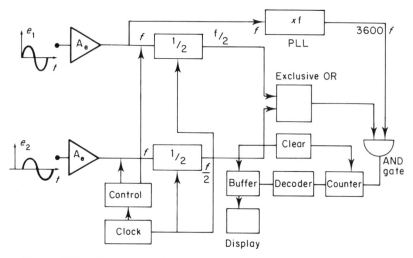

Figure 5.78 Phase measuring system using a PLL frequency multiplier

The circuit of Figure 5.78 includes a device for division by 2 in both input channels. This permits signals as long as the period to be measured, so that the phase can be converted from 0 to 360°. The signals coming from those two dividers are applied to an exclusive-OR gate which generates a θ long pulse on each period. If we need to know how late e_2 is compared to e_1, both dividers need to be controlled. In the case of positive edge dividers, the divider of the second input will be triggered after the rising edge of e_1 to measure the time separating e_2 from e_1.

In case of a 0.1° resolution, the frequency multiplier must generate a signal with a frequency 3600 times higher than the input signal frequency in which case the number of pulses recorded by the counter during interval θ will give an immediate reading of the phase difference.

The counter may be BCD type with a decoder and a buffer used for digital display. The negative edge of signal θ controls the clear input of the counter and the display of the buffered number. A very good accuracy is obtained for low frequencies, but it is difficult to measure high frequencies.

Numerical example At 10 kHz, a period lasts 100 μs. An increment of 0.1° corresponds to 10 μs/3600 = 30 ns. 30 ns is the rise-time of most TTL devices. We have to choose between a faster logic with a higher price for the system, and frequency multiplication of 360, which reduces the accuracy to one impulse per degree.

It can be seen that the action of the VCO is comparable with an integrator in the feedback loop when the PLL is considered in servo-theory. If the input frequency is not stable, the output frequency will vary. The display will show erratic values as long as the loop is not locked. The loop will lock only if the input frequency is stable enough. Another cause of error comes from the validity of the counting. The uncertainty on the measurement of the number of impulses may be ±1 depending on the VCO phase.

Numerical example A three digit display produces a result which is 0.1° degree short (3600 frequency multiplier). At full scale (360°), the accuracy is estimated at 0.3%.

5.7.7.2 Low disturbance flowmeter

The flowmeter is shown in Figure 5.81. Assume the liquid to be water. An ultrasonic wave is produced in A, $E(t) = E_0 \cos 2\pi f_0 t$; the receiver collects in A' a signal expressed by:

$$s(t) = KE_0 \cos\left[2\pi f_0 t - \phi_0 - \frac{2\pi \Delta d}{\lambda_0}\right].$$

Δd is dependent on time, flow in the pipe and pipe diameter.

$$\phi_0 = \frac{2}{\lambda_{\text{pipe}}} 2e + \frac{2\pi}{\lambda_0} d$$

with: d = pipe diameter
e = pipe thickness
λ_{pipe} = ultrasonic wavelength in the pipe
λ_0 = ultrasonic wavelength in the fluid.

$s(t)$ is analysed through a correlator which computs the integral

$$C_{\text{ss}}(\tau) = \frac{1}{T} \int_0^T s(t)s(t - \tau)\, dt.$$

$s(t)$ is supposed to be stationary.

An increasing flow of water in the pipe induces a decrease of τ_m in the autocorrelation function.

The equation describing $C_{\text{ss}}(\tau)$ is approximately

$$C_{\text{ss}}(\tau) = \frac{\sin 2\pi f_m \tau}{2\pi f_m \tau} \quad \text{with} \quad f_m = \frac{1}{\tau_m}.$$

This function is interpreted in the frequency domain by Figure 5.79. An increase in the speed of water is expressed by widening of the spectrum of $s(t)$. Consequently, the first zero crossing of the autocorrelation function allows a calculation of the flow in the pipe.

The use of a phase-locked loop allows us to keep

$$\phi_0 = \pm(n + 1)\frac{\pi}{2}$$

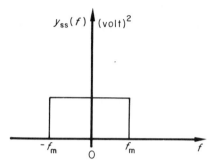

Figure 5.79 The spectrum width of $s(t)$ depends on the water velocity

in order to increase the signal sensitivity of $s_2(t)$ concerning the fluctuation term $\Delta\phi = 2\pi(\Delta d/\lambda_0)$. When there is no flow in the pipe signal $s(t)$ is equal to

$$s(t) = KE_0 \cos(2\pi f_0 t - \phi_0).$$

When the liquid is flowing through the pipe the term $2\pi\Delta d/\lambda_0$ appears in the phase of signal $s(t)$.

The low-pass filter suppresses the frequency $2F_0$ at the phase comparator output. The PLL follows the low-level frequency fluctuations caused by $2\pi(\Delta d/\lambda_0)$.

Systems using the autocorrelation technique are very critical to operate. On the one hand, the measurement depends on water purity and temperature; on the other hand, a recalibration is required for each new pipe diameter.

That is why cross correlation is recommended in most instances. Its functional principle is described in Figure 5.82, which includes two ultrasonic ceramic transmitters and two receivers. The transmission frequency is equal to f_0. The cross correlation function between $s_1(t)$ and $s_2(t)$

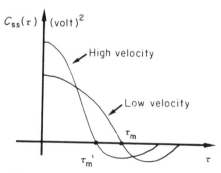

Figure 5.80 Flow in the pipe is determined by τ_m

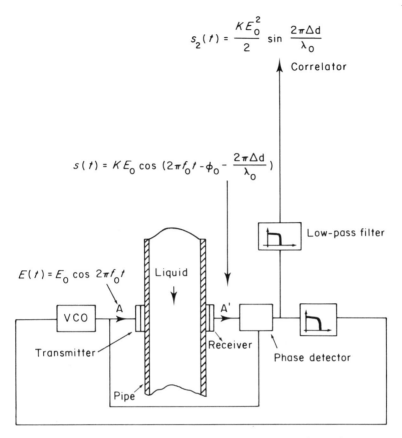

Figure 5.81 Flow-measuring system using PLL and correlator

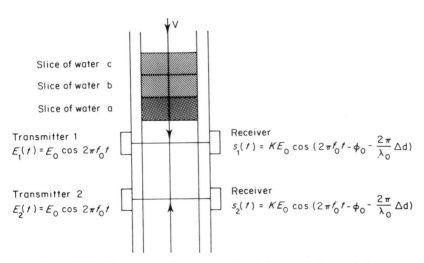

Figure 5.82 Flow-measuring sytem using intercorrelation technique

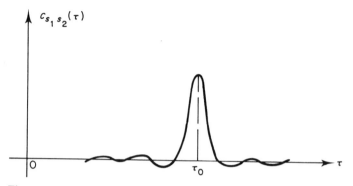

Figure 5.83 Representative curve of intercorrelation function between $s_1(t)$ and $s_2(t)$

can be expressed by

$$C_{s_1s_2}(\tau) = \frac{1}{T}\int_0^T s_1(t-\tau)s_2(t)\,dt.$$

In order to explain the functional principle of the flowmeter, water in the pipe is divided into slices whose thickness is equal to the ceramic transducer diameter (Figure 5.82). When the water slice a flows through the first transmitter–receiver system, it produces a phase modulation of $s_1(t)$, whereas, when the water slice b flows through the same system, it produces another phase modulation of $s_1(t)$ which is not identical with the preceding one. The phase modulation of $s_1(t)$ and $s_2(t)$ will be the same only when the same water slice flows through two transmitter–receiver systems. This effect can be expressed by the function $C_{s_1s_2}(\tau)$ as shown in Figure 5.83: τ_0 corresponds to the duration for the same water slice going from the first transmitter–receiver to the other one. Assuming that the distance between two transmitter–receiver couples, equal to 1: τ_0, can be measured by the correlator, then we can obtain the fluid velocity

$$V_m = \frac{1}{\tau_0};$$

so that it is now possible to calculate the flow in the pipe. Signal treatment can be carried out as in the preceding method.

It is interesting to mention that most industrial flowmeters are designed using the Doppler effect or the influence of liquid velocity on the speed of signal propagation in the pipe (Fizeau flowmeter).

PART 2

Electronics Associated with Digital Measuring Systems

Chapter 6
COMPARATORS AND ANALOGUE SWITCHES

6.1 COMPARATORS

Comparators frequently serve as the interface between analogue and digital systems. A voltage comparator senses the relative polarity of the difference between the voltages applied to the comparator's two inputs. Its output presents two states of binary voltage level: logic 0 and logic 1.

Most comparators are essentially high-gain differential amplifiers operating in the open-loop manner and driving an output stage conditioned for specific logical voltage levels. In some cases op amps can serve as comparators by clipping their output voltage to the required logic levels. However, since op amps are designed to preserve a linear relationship between input and output signals, they frequently have response times which may be too slow for many comparator applications.

6.1.2 Basic Circuits

Figure 6.1 shows an operational amplifier without feedback representing the basic circuit of a comparator and its transfer characteristic.

The output of an ideal voltage comparator will be at a level corresponding to logic 1 whenever the voltage difference existing between the noninverting and inverting inputs is positive, or

$$V_{out} = \text{logic 1 when } \varepsilon = V_+ - V_- > 0.$$

When this voltage difference is negative, the output is a logic 0, or

$$V_{out} = \text{logic 0 when } \varepsilon = V_+ - V_- < 0.$$

The output of an ideal comparator changes state when $V_+ = V_-$. In practical comparators, the output does not immediately reach the saturation level at zero crossing of the input difference, because the transition is limited by the slew rate of the op amp. For frequency-compensated standard operational amplifiers, it is about $1\,V/\mu s$. A rise from $-12\,V$ to $+12\,V$ therefore takes $24\,\mu s$. An additional delay is incurred due to the

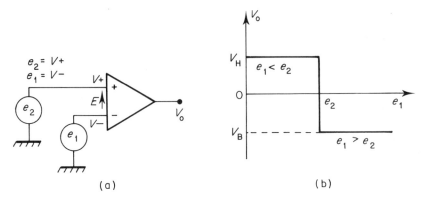

Figure 6.1 Comparator: (a) basic circuit; (b) transfer characteristic

recovery time needed for the op amp to recover from the saturation region.

In comparator applications, frequency compensation is unnecessary and the feedback omission can improve the slew rate and recovery time by a factor of about 20. However, in that case any stray coupling between the output and the input terminals of the comparator can occasionally cause oscillations.

The available range of input voltages for a comparator, as in Figure 6.1, is limited. If large voltages are to be compared, the comparators can be used in the circuit shown in Figure 6.2. The comparator changes its output when ε crosses zero. This is the case when

$$\frac{e_1}{R_1} = -\frac{e_2}{R_2}.$$

The voltages to be compared must therefore have different polarities.

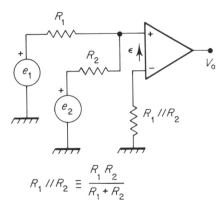

Figure 6.2 The comparator changes its output when ε crosses zero

Protective diodes can be added to further limit the differential input voltage range to values between ±0.6 V. Among factors which affect the ideal operation of a practical voltage comparator, one can mention the slew rate, recovery time, offset voltage, bias current, common-mode rejection. All these performance factors must be taken into account for the correct choice of a comparator in a given application.

6.1.3 Characteristics

6.1.3.1 Accuracy

Comparator accuracy is mainly affected by errors due to the input offset voltage, the limited voltage gain of the comparator, the input-bias current, and the finite common-mode rejection.

(1) Input offset voltage is a measure of the magnitude of the actual difference between positive and negative inputs which causes the output voltage to indicate a state change. This parameter depends on the logic supply level voltage and the comparator's load.

(2) Finite voltage gain (A) affects the switch-over voltage, i.e. the input voltage at which the output voltage changes from high logic state to low logic state. As a result, the zero-crossing point of the output voltage is shifted from $V_+ = V_-$ to

$$\varepsilon = V_+ - V_- = e_2 - e_1 = \frac{V_0}{A}.$$

This error is usually negligible for instance, with a voltage gain of 100.000, the uncertainty range is very small and corresponds to $30\,\mu$V for a 3 V output swing.

(3) Input bias current is the average of the two base biasing currents for the buffering transistors, connected to the differential input stage. This factor affects the switch-over voltage when the signal source has a high internal impedance. In the case of Figure 6.2, the input bias current i generates an additional input offset voltage equal to $iR/2$ with $R_1 = R_2 = R$. Although input bias current and input offset voltage effects can be eliminated by external offset adjustment compensation, it is difficult to compensate their drift effects entirely.

(4) Common-mode voltage given in the specification could be considered as a limit value not to be exceeded. For instance in Figure 6.1, the common mode voltage $(e_1 + e_2)/2$ contributes to the input offset voltage in a direct manner.

6.1.3.2 Response time

One important index of the dynamic switching characteristics of the voltage comparator is the response time. It is specified by means of response

Table 6.1 Some examples are given in Table 6.1

Model	Offset voltage	Bias current	Response time
Accurate: LM 139 National Semiconductor	2 mV	5 nA	1.3 µs
Fast: µA 760 Fairchild	6 mV	7.5 µA	16 ns

diagrams on the device's specification sheets. The response is strongly influenced by the input-signal test conditions, especially with the amount of voltage overdrive applied on the input signal. Usually, the response times quoted are the times necessary for the outputs to change state when a 100 mV input signal with a 5 mV overdrive is applied.

6.1.3.3 Output voltage levels

The output voltages from a 'simple' comparator are functions of the saturation voltage and the supply voltage. If a circuit is to provide given output levels, the connections of Figure 6.3 can be used, where the output levels are set by diode voltages.

When the output voltage reaches the values $\pm(0.6\,\text{V} + V_Z)$, the Zener diodes effect feedback for the operational amplifier and in this way a further rise in voltage is prevented. As the amplifier does not saturate, there is no recovery time.

Usually, the output stage of a comparator is designed to be very flexible in the provision of voltage levels which are compatible with most logic types e.g. RTL, DTL, TTL and MOS logic. The model LM 339 from National

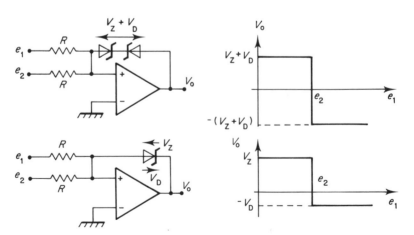

Figure 6.3 The comparator output level can be defined by zener diode voltages

Semiconductor, for example, has a response time of 500 ns and an open collector output to which external connections could be made to select appropriate logic levels. The Signetics model ICNE 521 includes a TTL compatible output with a response time of 8 ns. A particularly short delay of 5 ns is attained with Advanced Micro Devices comparator Am 685 which has an ECL compatible output.

6.1.4 Window Comparator

A window comparator can determine whether or not the value of the input voltage lies between two reference voltage levels. This requires that the output signals of two comparators are connected to a logic circuit, as in Figure 6.4. The integrated circuit NE 521 mentioned previously is particularly well suited as it contains on a single chip not only two comparators with level-shift circuitry but also two NAND gates. As can be seen in Figure 6.4, the output is at boolean 1 only if

$$e_1 < e_i < e_2$$

as then both comparators outputs are also at boolean 1.

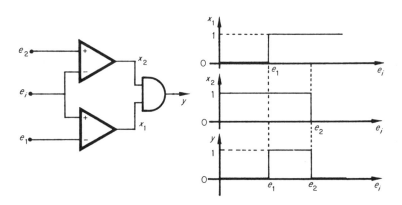

Figure 6.4 Window comparator

6.1.5 Application Considerations

As indicated earlier, a comparator is generally a high-gain, wide-band amplifier operating in an open-loop configuration, without frequency compensation, so that any capacitive coupling between the input and output may lead to oscillation. Careful wiring and a low-impedance signal source are recommended to reduce stray capacitances effects at the input.

6.1.6 Hysteresis Comparators

The noise levels riding on low-level input signals can be particularly troublesome in level detectors. For example, when the input signal approaches the reference level, noise can cause the comparator to change state incorrectly. If the amplitude of the input signal is exactly equal to the reference voltage, the comparator will change state randomly in response to the polarity of the noise voltage. To avoid this difficulty, positive feedback is frequently employed to alter the comparator's transfer characteristic. The modified circuit, known as a Schmitt trigger, produces two voltages V_2 and V_1, which are the levels at which the comparator changes state. The difference between V_2 and V_1 is the hysteresis voltage, which is made somewhat larger than the maximum expected noise voltage.

6.1.6.1 Inverting Schmitt trigger

In the Schmitt trigger of Figure 6.5, the hysteresis is effected by a positive feedback around the comparator using the voltage divider R_1, R_2. If a large negative voltage e_i is applied, the output voltage is equal to $V_0 = S_M$ and then we have

$$V^+ = S_M \frac{R_1}{R_1 + R_2} V_2.$$

If the input voltage is changed towards positive values, V_0 does not change at first. Only when e_i reaches the value V_2, due to positive feedback, does V_0 fall very quickly to the value S_m. The switching speed equals the slew rate of the amplifier. The potential V_+ assumes the value

$$V_1 = \frac{R_1}{R_1 + R_2} S_m$$

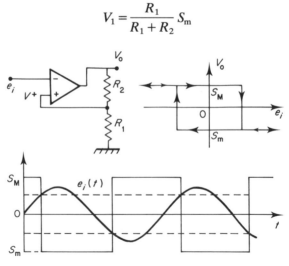

Figure 6.5 Hysteresis comparator: inverting Schmitt trigger

Figure 6.5 shows an important application of the Schmitt trigger. This circuit converts an input voltage of any shape into a binary-wave output voltage. This has a fixed rise time which is independent of the shape of the input signal.

6.1.6.2 Non-inverting Schmitt trigger

If one of the two inputs of the comparator in Figure 6.2 is connected to the output, the result is the non-inverting Schmitt trigger of Figure 6.6. The hysteresis voltage is

$$V_H = \frac{R_1}{R_2}(S_M - S_m).$$

In the same way that an inverting amplifier can be expanded to become a summing amplifier, so can the Schmitt trigger be extended to become a summing Schmitt trigger. For this purpose additional resistors are connected to the (+) input providing further voltage inputs. This method is shown in Figure 6.7. The trigger levels can be varied by means of the voltage E_{ref},

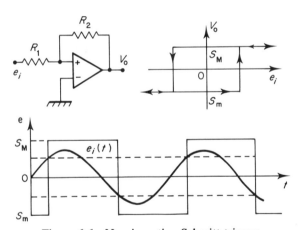

Figure 6.6 Non inverting Schmitt trigger

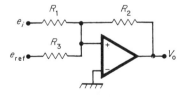

Figure 6.7 Summing Schmitt trigger

but the hysteresis remains unchanged.

Threshold for switch-on: $e_i = -\dfrac{R_1}{R_2} S_m - \dfrac{R_1}{R_3} E_{\text{ref}}$

Threshold for switch-off: $e_i = -\dfrac{R_1}{R_2} S_M - \dfrac{R_1}{R_3} E_{\text{ref}}$

6.1.7 Precision Schmitt Trigger

The trigger levels of the Schmitt triggers described so far are not very accurate with operational amplifier circuitry. The accuracy can be improved by connecting a comparator to an analogue switch, as in Figure 6.8. Thus, depending on the state of the comparator, the output voltage assumes two precisely defined values, V_1 or V_2. Due to the positive feedback via the voltage divider $R_1 R_2$, the trigger levels are given as

$$e_1 = \frac{R_1}{R_1 + R_2} V_2, \qquad e_2 = \frac{R_1}{R_1 + R_2} V_1.$$

They are therefore no longer dependent on the level at which the

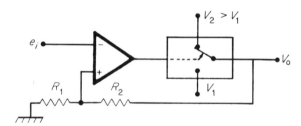

Figure 6.8 Precision Schmitt trigger with analog switch

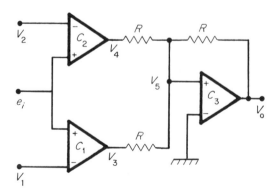

Figure 6.9 Precision Schmitt trigger with two comparators

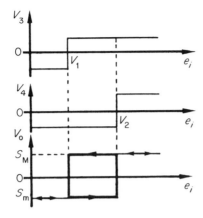

Figure 6.10 Transfer characteristic

operational amplifier saturates, as is the case for the circuit in Figure 6.5. Often a high accuracy is required for the trigger levels but not for the output voltage. This demand can be fulfilled without the use of an analogue switch, as Figure 6.9 shows. The amplifier C_3 is operated as a summing Schmitt trigger, and two comparators C_1 and C_2 accurately specify the trigger levels.

If the input voltage is larger than the two reference voltages V_1 and V_2, then $V_0 = S_M$. When the input voltage falls below the larger of the two reference voltages, V_0 does not change. This is because one of the two output potentials, V_3 and V_4, is at S_M and the other at $S_m \simeq -S_M$. Hence $V_5 \simeq \frac{1}{3}S_M > 0$. Potential V_5 becomes negative only when the input voltage has also fallen below the second reference voltage. At this instant, the output jumps from S_M to S_m. The trigger level for the S_M state is therefore identical to the smaller of the two reference voltages, and the trigger level for the S_M state identical to the larger. This is illustrated by the transfer characteristic in Figure 6.10. If, instead of the operational amplifiers C_1 and C_2, special comparators having level-shift circuits are used, the amplifier C_3 can be replaced by an RS flip-flop, as shown in Figure 6.11. The flip-flop is set

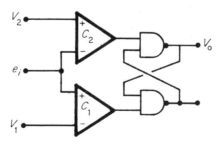

Figure 6.11 Precision Schmitt trigger with RS flip-flop

when the input voltage V_2 exceeds the trigger level V_1, and is reset when it falls below the value V_1.

The IC NE 521 mentioned previously consists of two comparators, each with a NAND gate connected to it. By an appropriate external connection, the desired function can be implemented by a single IC.

6.2 ANALOGUE SWITCHES

One of the most common control elements in electrical circuitry is the simple on–off switch. This has evolved over the years from the manually operated circuit breaker of the early experimenters to the multiswitch integrated circuit of today. In every application, the function of the switch remains the same, i.e., to isolate or connect two sections of an electrical circuit.

With the increasing demands of the modern circuits of today, it has become evident that electromechanical switches alone cannot meet all requirements and that there are applications in which only electronic types are viable. By far the most popular of these is the semiconductor switch.

In recent years, semiconductor switches have been introduced into application areas that used to be the exclusive domain of electromechanical devices. Solid-state switches are now used in sample-and-hold circuits, multiplexers, high-power switching, chopper circuits, etc., whereas in the past some form of electromechanical switch would have been used. In comparison with electromechanical types, semiconductor devices exhibit greater resistance to adverse environments, consume less power and are much smaller in size and weight.

Figure 6.13 shows the equivalent circuit of a transistor switch at its two operating modes: on and off. In the off state, a cut-off transistor behaves like a circuit formed by a high-value resistance (R_{off}) shunted by a junction capacitance (C_{off}). In the on state, the saturated transistor is represented by a low resistance (R_{on}). In addition to these imperfections, there are other limitations such as the breakdown voltage, the maximum current and the maximum switching speed.

Field effect transistors present a high off-to-on resistance ratio (about 10^9), a relatively simple equivalent on circuit (resistive and bilateral), inherent high-speed switching capability, and if properly used, a high degree of control-signal isolation. Unlike the bipolar transistor, which exhibits a

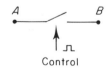

Figure 6.12 Analog switch

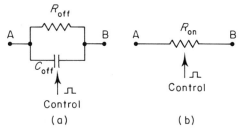

Figure 6.13 Simplified models of analogue switch: (a) OFF state; (b) ON state

significant V_{CE} offset voltage, the V_{DS} offset for a FET is usually negligible. FET switches are, therefore, ideal components for use in switching applications.

6.2.1 Junction FET Switch

Suppose the input voltage varies between V_1 and V_2 ($V_1 > V_2$). Figure 6.14 shows a JFET used as a series switch. This JFET is off and the output voltage is zero if the control voltage V_c is made more negative than the maximum negative input voltage and the difference is at least equal to the pinch-off voltage, V_p, of the JFET.

The JFET switch is on if $V_{GS} = 0$, and off if $V_G < V_2 - V_p$. The condition $V_{GS} = 0$ to make the FET conduct is not so easy to fulfill as the source potential is not constant. A solution to this problem is shown in Figure 6.15 where diode D becomes reverse-biased if V_c is made larger than the most positive input voltage, and therefore $V_{GS} = 0$ as required.

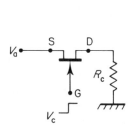

Figure 6.14 JFET analogue switch

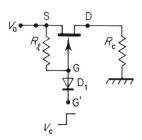

Figure 6.15 JFET is protected by diode $D1$ during its ON state

---

Numerical Example Suppose that $-10\,\text{V} < V_a < +10\,\text{V}$ and $V_p = 6\,\text{V}$. The command voltage V_c for the switching circuit shown in Figure 6.15 is therefore

— to cut the JFET off: $V_G < V_2 - V_p$ with $V_2 = -10\,\text{Volts}$
$$V_G < -10 - 6$$
to give an adequate design margin let $V_G = -18\,\text{V}$

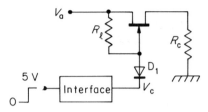

Figure 6.16 TTL compatible interface circuitry

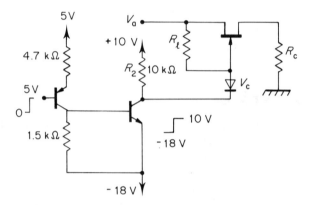

Figure 6.17 TTL compatible interface circuitry

— to make the JFET conduct: $V_{GS} = V_G - V_S = 0$. As discussed previously, this condition is satisfied if $V'_G = 10$ V

Generally, the command voltage V_c is generated by an interface circuit which is compatible with standard logic families such as TTL, CMOS, DTL or RTL. Figures 6.16 and 6.17 show one example of TTL compatible interface circuitry.

In integrated circuits a hybrid structure is frequently used, i.e. the resistor R_2 in Figure 6.17 may be replaced by a JFET connected as a resistance (example of industrial realizations are D111F, D112F, D113F from Siliconix). The interface circuitry, in some devices, is integrated on the switch chip (for example as in the National Semiconductor AH 211H).

The driver circuit discussed above is not necessary in current-switching application as shown in Figure 6.18. In that case, a p-channel JFET with pinch-off voltage less than $+5$ V is generally used so that the gate can be directly connected to the TTL signals. Figure 6.18 shows a p-channel JFET connected as a series switch to the summing point of an op amp which converts the drain switch current to output voltage. Hence, the drain potential V_D is fixed and equals zero.

The series resistance R_1 is chosen such that the voltage across the conducting FET remains small, i.e. $V_D \simeq V_S \simeq 0$. Therefore, the FET is

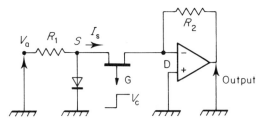

Figure 6.18 P channel JFET connected as a series switch to the summing point of an op amp

switched to on when $V_c = 0$, logic 0, independently of the magnitude of the input voltage. In that case, the current I_s driving through the FET is given by

$$I_s = \frac{V_a}{R_1 + R_{ON}}$$

and the op amp output voltage is then

$$-\frac{V_a}{R_1 + R_{ON}} R_2.$$

The FET is switched to off when $V_C = V_G = +5\,\text{V}$, logic 1, since the condition $V_{GS} > V_p$ holds for any values of the input signal V_a. It is quite obvious for $V_a < 0$ that $V_{GS} = 5 - V_a > 5 > V_p$. For $V_a > 0$, the source potential V_S rises and the diode becomes forward biased. In this way, V_S is limited to values less than $+0.6\,\text{V}$ and $V_{GS} > 4.4\,\text{V} > V_p$.

To reduce the dependence of the output voltage on R_{ON}, the manufacturer usually includes, in the feedback loop in series with resistance R_2, a compensating JFET operating as a resistance (Figures 6.19 and 6.20). As JFET R_{ON} resistances and their temperature drift within a package are approximately the same, the overall system gain is R_2/R_1.

With this type of switch, a high value of current I_S is recommended. However, as I_S increases, the leakage current I_{GON} also increases (Figure

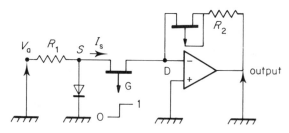

Figure 6.19 Compensating JFET operates as a resistance

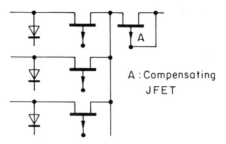

Figure 6.20 Structure of an analogue switch module with a compensating JFET

6.21). Besides, as I_S approaches the saturation current I_{DSS} ($I_D = I_{DSS}$ when $V_{GS} = 0$), resistance R_{ON} leaves the linear region.

A practical rule consists of maintaining $I_S \leq \dfrac{I_{DSS}}{10}$.

It follows that R_1 must be greater than $\dfrac{V_{amax}}{I_{DSS/10}}$.

Numerical Example
$V_a = 10\,\text{V} \quad I_{DSS} = 20\,\text{mA}$
$$R_{1\,\text{minimum}} \geq \frac{10\,\text{V}}{20\,\text{mA}/10} = 5\,\text{k}\Omega$$
The current I_S is then: $\dfrac{10\,\text{V}}{5\,\text{k}\Omega} = 2\,\text{mA}$.

At this value of I_S, the model AM 9710 CN of National Semiconductor has a negligible leakage current I_{GON} (less than 5 nA at 85°C) as indicated in its specifications. In comparison, the leakage currents of other analogue switches can exceed several tens of microamperes (μA).

Another error source illustrated in Figure 6.21 is the sum of leakage currents coming from other OFF JFET switches: I_{DOFF}. At normal temperature, this current is negligible against op amp imperfections such as input bias current and offset voltage as its typical values are in the 10–20 pA range but they can exceed several nA at 85°C.

6.2.2 MOSFET Switches

The MOSFET or metal-oxide–semiconductor FET uses a film of high purity dielectric such as silicon dioxide or silicon nitride to insulate the gate from the channel. Due to the insulation properties of the MOS gate, both positive and negative gate voltages may be applied to unprotected MOSFETs. Four forms of MOSFET are possible:
  (a) *n*-channel depletion.   (b) *p*-channel depletion
  (c) *n*-channel enhancement   (d) *p*-channel enhancement

The depletion MOSFET differs from the enhancement by the presence of an initial channel between source and drain. The device can therefore be operated in both depletion and enhancement mode. Otherwise a negative (for n-channel MOS) or positive (for p-channel MOS) gate voltage must be applied to turn the channel off so that in this sense the device behaves exactly as an n- or p-channel JFET respectively. The design and operation of an n-channel enhancement MOSFET is similar to the p-channel enhancement MOSFET except for reversal of voltage polarities, so only the p-channel enhancement (also called PMOS) will be reviewed in detail.

An indealized cross-section through a p-channel enhancement mode MOSFET is illustrated in Figure 6.22. The device consists of an n-type substrate into which a p-type impurity is diffused to form separate sources and drains. The metal gate is insulated from the substrate by an oxide layer, to avoid destructive currents flowing from source and drain into the substrate. As the substrate voltage must be sufficiently high to reverse bias these two p–n junctions it is usually connected to the highest potential available.

The device is normally off. As the PMOS gate is forced negative with respect to the source, the resultant field attracts holes or p-type carriers, forming a conductive p-channel between source and drain. The point where conduction begins is called the threshold voltage V_T. As the gate-to-source voltage V_{GS} is forced more and more negative, the conductive channel widens and the drain-to-source resistance R_{DS} decreases. $|V_T|$ is commonly 3–4 V. However, for a useful on resistance of 200 ohms, it is necessary to force V_{GS} to -10 V.

In the unsaturated region, i.e. $|V_{DS}| < |V_{GS}| - V_T|$, the drain current is approximated by

$$I_D = \beta \left\{ V_{DS} \left[V_{GS} + |V_T| \right] - \frac{V_{DS}^2}{2} \right\}$$

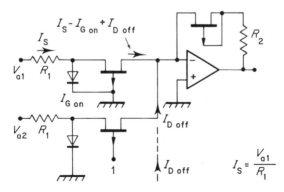

Figure 6.21 Influence of leakage currents $I_{G on}$ and $I_{D off}$

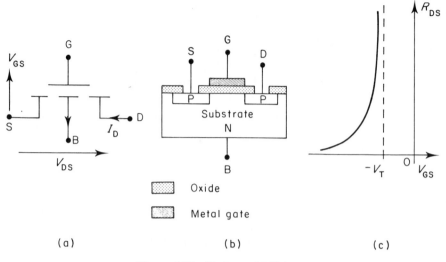

Figure 6.22 P channel MOSFET

Thus the on resistance is given by:

$$R_{DS} = \frac{V_{DS}}{I_D} = -\frac{1}{\beta}\frac{1}{(V_{GS}+|V_T|-\frac{1}{2}|V_{DS}|)} \simeq \frac{-1/\beta}{V_{GS}+|V_T|}.$$

for low values of V_{DS}, β is the device constant function of the geometry of the MOSFET.

As with the JFET, the R_{DS} of a MOSFET is dependent not only on V_{GS} but also upon V_{DS}, the minimum on resistance occurring when $|V_{GS}-\frac{1}{2}V_{DS}|$ is a maximum. Therefore, if MOS devices are used in switching circuits that have a fixed gate voltage applied in the on state, the channel resistance will be modulated by any variation in the analogue signal voltage.

Figure 6.23 shows the equivalent circuits of a MOSFET in the on and off states.

Since the MOSFET has an extremely thin induced channel, it tends to have a higher on resistance (100 ohms) than a junction FET (25 ohms) of similar size. The effects of temperature on channel resistance are similar in both MOS and junction FETs. Both MOS and junction FETs exhibit a positive temperature coefficient of resistance typically +0.4%/°C for MOS and +0.75%/°C for JFET.

In both types, the two most important capacitances are the gate-to-source capacitance and the gate-to-drain capacitance for JFET or the body-to-source capacitance and the body-to-drain capacitance in the case of a MOSFET. Their values are about 5–10 pF.

The most important leakage current is I_{Doff}. As it flows into the load, it thus generates an offset voltage. Although this current is negligible

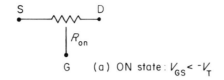

(a) ON state: $V_{GS} < -V_T$

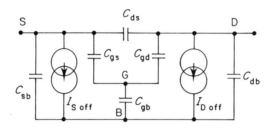

(b) OFF state: $V_{GS} > -V_T$

Figure 6.23 Models of MOSFET switch in the ON and OFF state

(nanoamperes) at normal temperature it increases rapidly with the temperature.

Among solid-state switches, the MOSFET switch is the simplest to use. The switch itself is easy to construct, and the on/off drive circuit is quite basic. However, the extremely high gate-to-channel resistances (in excess of 10^{15} ohms) present a problem when handling MOSFETs. Electrostatic charge build-up at the gate can cause the gate-channel capacitance to charge up to voltages which can exceed the dielectric breakdown of the gate-to-channel (about 120 V). Permanent damage to the dielectric (oxide film) can

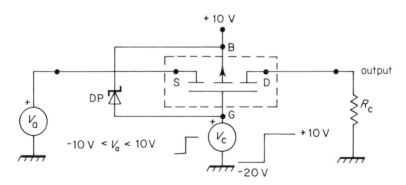

DP: Protection diode

Figure 6.24 Control signal can be directly applied on the gate of MOSFET switch

therefore occur. To protect this dielectric, many MOSFETs are manufactured with an integrated Zener clamp.

Numerical Example
Switching voltage: $-10\,V < V_a < +10\,V$
Command voltage: $\leftarrow +10\,V$ (OFF)
(ON) $-20\,V \rightarrow$
Substrate potential: $+10\,V$.
Off state: $V_G = +10\,V$
if $V_a = -10\,V$, then $V_{GS} = 10 - (-10) = 20\,V$
if $V_a = +10\,V$, then $V_{GS} = 10 - 10 = 0\,V$.
Therefore when $-10 < V_a < +10$: $V_{GS} > -V_T$ (for V_T is positive for a PMOS).
On state: $V_G = -20\,V$
if $V_a = -10\,V$, then $V_{GS} = -20 + 10 = -10\,V$
if $V_a = +10\,V$, then $V_{GS} = -20 - 10 = -30\,V$
PMOS is turned on if its threshold voltage V_T is less than 10 V.

Two basic types of PMOS drive gate are shown in Figures 6.25 and 6.26. The driver circuit shown in Figure 6.25 is the same as the one shown in Figure 6.17 used for the JFET, except that diode D and resistor R_l are no longer necessary because there is no current flow from gate to channel due to the oxide layer, even for control voltage larger than the most positive input voltage.

The gate is then direct driven by a level shifting circuit that also serves as

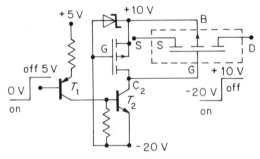

Figure 6.25 Hybrid interface circuit

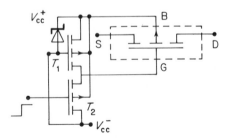

Figure 6.26 MOSFET interface circuit

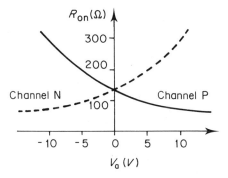

Figure 6.27 R_{ON} vs. analogue voltage V_a

an amplifier, since it converts TTL logic signals (0 V and 5 V) into the −20 V to +10 V gate signals required to turn the MOSFET ON or OFF. This range of drive gate amplitude permits switching an analogue signal of ±10 V.

A logical extension of the circuit shown in Fig. 6.25 is the one shown in Fig. 6.26. This is an all PMOS arrangement which lends itself to large monolithic IC arrays of switches.

The chief disadvantage of the PMOS gate is the 3:1 ratio in R_{DS} for analogue voltages varying between −10 V and +10 V (Figure 6.27). An ideal means of avoiding this variation of R_{DS} is to use a CMOS gate. It uses two enhancement mode MOSFETs in parallel (one PMOS and one NMOS) between each source and drain terminal (Fig. 6.28). To turn the NMOS FET on, V_{GS} is made more positive than V_T, which is now +3V to +4V. Therefore, the CMOS switch is turned on by driving the PMOS gate negative and the NMOS gate positive simultaneously (Fig. 6.28).

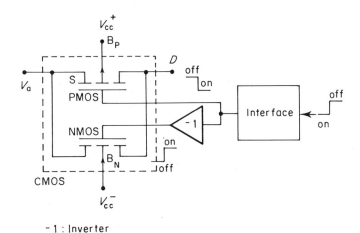

−1 : Inverter

Figure 6.28 CMOS switch control

Numerical Example Assume that the interface circuit converts TTL logic signals into -15 V and $+15$ V. These potentials are then applied to a digital CMOS inverter which commands the CMOS switch.

When the input inverter is -15 V:

$$V_G(P) = -15 \text{ V}$$
$$V_G(N) = +15 \text{ V},$$

and when the input analogue voltage swings from -15 V to $+15$ V, at least one of the two switches is on.

PMOS is on when $V_a = +15$ V for $V_{GS}^{(P)} = -15 - 15 < -V_{TP}$
NMOS is on when $V_a = -15$ V for $V_{GS}^{(N)} = 15 + 15 > V_{TN}$

When the input inverter is $+15$ V:

$$V_G(P) = 15 \text{ V}$$
$$V_G(N) = -15 \text{ V}$$

and both switches are off. For $-15 < V_a < +15$:

$$V_{GS}(P) = 15 - V_a > 0 > -V_{TP}$$
$$V_{GS}(N) = -15 - V_a < 0 < V_{TN}.$$

The curves of Figure 6.29 illustrate the variation of channel resistance with applied switch voltage V_a. While the resistance of the n-channel increases with positive voltage, and the resistance of the p-channel increases with negative voltage, their parallel combination is relatively constant $\pm 20\%$ as opposed to PMOS, which changes as much as 600% over the entire analogue signal range.

One of the major advantages of CMOS apart from the constancy of the on resistance is that the analogue signal can equal the voltage across the supply rails (± 15 V, for example). This is unique to CMOS.

Figure 6.30 shows a CMOS and its inverter circuit. The driver stage is

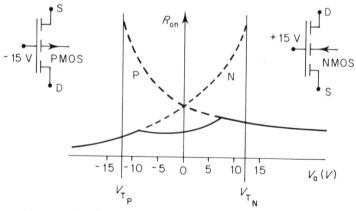

Figure 6.29 R_{on} vs. analogue voltage V_a for CMOS switch

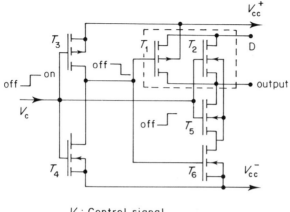

V_c: Control signal

Figure 6.30 CMOS switch is driven by complementary signal

basically simple since the PMOS and NMOS switches (T_1, T_2) are driven by complementary signals from the input and output modes of an inverter (T_3, T_4).

In practice, the analogue switch circuit is slightly complicated by the presence of 'body snatching' transistors T_5 and T_6 as shown in Figure 6.30. T_5 and T_6 switch the body of T_2 to either the negative supply (in the off state) or the source (when the switch is on). The result of a positive clamping of the body to the source is a low on threshold voltage which is not modulated by the analogue voltage. In the off condition the isolation and breakdown voltage are high and the leakage is low when the body is clamped to the negative supply.

Examples of industrial realizations
(1) The model DG 172 CJ of Siliconix is a monolithic circuit consisting of four PMOS switches with built-in driver which accepts low command signal voltage (0.8–2 V). All gates are protected from static charge build-up by an incorporated Zener diode connected between the four gates inputs and the device body.
(2) The Analog Devices AD 7510, AD 7511 consist of four independent SPST analogue switches packaged in a 16-pin DIP. They differ only in that the digital control logic is inverted.

Note that monolithic modules generally include several switches and permit varied operating configurations. These configurations are determined by the letters S (for single), D (double), P (pole), T (throw) whose associations indicate the module function.
Examples
(a) HI 5051 Intersil, SPDT
(b) HI 5049 Intersil, DPST

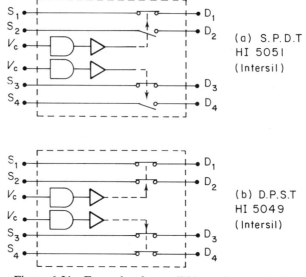

Figure 6.31 Example of monolithic analogue switch

6.2.3 Diode Switch

Diodes are also suitable for use as switches because of their low forward resistance and high blocking resistance. If a positive control voltage is applied to the circuit in Figure 6.32, the diodes D_5 and D_6 become reverse-biased. The impressed current I then flows through the branches D_1, D_4 and D_2, D_3 from one current source to the other. The potentials V_1 and

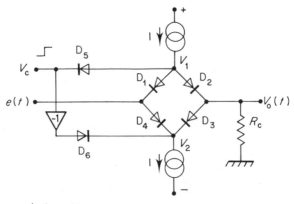

−1: Inverter

Figure 6.32 Diodes switch

V_2 thereby take on the values:

$$V_1 = e + V_D, \qquad V_2 = e - V_D.$$

The output voltage then is

$$V_0 = V_1 - V_D = V_2 + V_D = e$$

if the forward voltages V_D are the same. An offset voltage arises if this is not the case.

If the control voltage is negative, the two diodes D_5 and D_6 are forward biased. Potential V_1 takes on a high negative value and V_2 a high positive value. As can be seen in Figure 6.32, all diodes of the quartet D_1 to D_4 are then reversed biased. The output is separated from the input, and the output voltage is zero.

To keep switching times short and capacitive transients low, Schottky diodes are usually used. In this way, extremely fast switches can be built, the switching time of which may be below 1 ns.

6.2.4 Bipolar Junction Transistor Switch

The bipolar transistors have a lower R_{ON} than FET transistors. They are largely used to provide rapid and accurate switching systems for DACs (Analog Devices AD 550).

Figure 6.33 shows the application of a transistor as a short-circuiting switch. When the transistor is saturated, the output voltage is short-circuited, and when the transistor is cut off, the output voltage follows the input.

Figure 6.34 shows the application of a bipolar transistor as a series switch. A negative control voltage must be applied to cut off the transistor. It must be more negative than the most negative value of the input voltage. To make the transistor conduct, a positive control voltage is applied, which must be larger than the input voltage. The collector–base junction is then forward-biased and the transistor operates as a switch in reverse region mode. The offset voltage V_{CE} can be very small. The output voltage follows the input with the disadvantage that the base current flows into the input

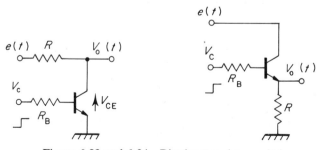

Figure 6.33 and 6.34 Bipolar transistor switch

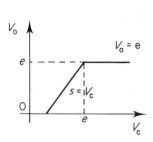

Figure 6.35 Transfer characteristic for positive input voltage

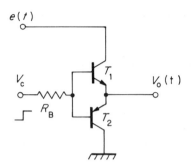

Figure 6.36 Series/short circuiting switch

voltage source. Thus large errors may occur unless the internal resistance of the source is kept very small.

The circuit in this mode of operation is known as a saturated emitter follower, since for control voltages between zero and e, it operates as an emitter follower for V_c. This is illustrated in Figure 6.35 by the transfer characteristics for positive input voltages.

If the saturated emitter follower is combined with the short-circuiting switch, a series/short-circuiting switch is obtained. Although a low offset voltage results for both modes of operation, it has the disadvantage that complementary control signals are required. The control is particularly simple if a complementary emitter follower is used, as in Figure 6.36. It is saturated in both the on and off mode if $V_{c\,max} > e$ and $V_{c\,min} < 0$. Due to the low output resistance, a fast switchover of the output voltage between zero and e is possible.

6.2.5 Sources of Error

Since FETs are good switches, the errors that they introduce are quite small. The main sources of these errors will now be investigated in order to understand the problems that could occur and hence enable a solution to be implemented if necessary. The errors can be divided into two groups:

(1) low-frequency or d.c. errors due to leakage currents and switch resistance;
(2) high-frequency errors due to device and stray capacitances.

6.2.5.1 Resistance error in the on state

The switch resistance is in series with the load and therefore causes some attenuation of the analogue signal. This can be a significant source of error if the load resistance is less than $100\,\mathrm{k\Omega}$ (Figure 6.37).

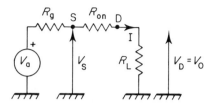

Figure 6.37 Error due to R_{on}

Referring to Figure 6.37,

$$V_0 = \frac{R_L V_a}{R_{ON} + R_L + R_g} \left(\simeq \frac{R_L}{R_{ON} + R_L} V_a \right).$$

Then, if we substitute some typical values such as: $V_a = 10$ V, $R_g = 50$ ohms, $R_{ON} = 100$ ohms, $R_L = 1$ kohm into the above equation, we have

$$V_0 = \frac{1000}{1150} V_a \text{ i.e. an } 8.7\% \text{ error.}$$

Or if $R_L = 100$ kΩ, then $V_0 = \frac{100\,000}{100\,150} V_a$, i.e. a 0.15% error. We note that while JFET R_{ON} is approximately constant, MOSFET R_{on} varies according to the effective V_{GS} generated by the analogue signal and in that case, distortion as well as attenuation of the analogue signal can occur.

6.2.5.2 Leakage error

In the on state, there will be leakage current flowing from the switch channel into either the driver circuit if the switch is a JFET, or into the body if the switch is a MOSFET. Typical values for $I_{D(ON)} + I_{S(ON)}$ are usually around 1–5 nA. Consequently their effect is insignificant compared to the analogue current which is normally several milliamperes.

In the off state, the main components of leakage current in a JFET are gate-to-source or gate-to-drain leakage. For a MOSFET, the leakage currents flow mainly from body to source or body to drain.

The effect of the off leakage can be assessed by considering a driver gate in the off state, having its source and drain connected as in Figure 6.38. The

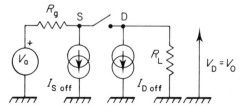

Figure 6.38 Error due to leakage current

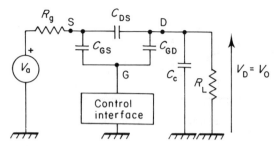

Figure 6.39 High-frequency equivalent circuit of a JFET switch and feedthrough due to switch capacitances

error voltage seen at the load will be $(I_{D(off)} \times R_L)$, which is $1\,\mu V$ for $R_L = 1\,k\Omega$ and $I_{D(off)} = 1\,nA$.

6.2.5.3 High-frequency off isolation

The high-frequency equivalent circuit of the JFET in the off state is given in Figure 6.39.

Since a typical value for C_{ds} is $0.1\,\mu F$, whereas C_{gs} and C_{gd} are usually at least an order of magnitude higher, it is apparent that if the gate were left floating the majority of the signal feed-through from source to drain would be via C_{gs} and C_{gd}. Therefore, if the drain-to-source off isolation at high frequency is to be maximized, the driver circuit should present as low a resistance as possible to a.c. ground, so that any signal coming via C_{gs} will

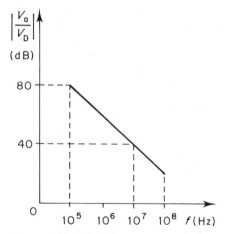

Figure 6.40 High-frequency equivalent circuit of a JFET switch and feedthrough due to switch capacitances

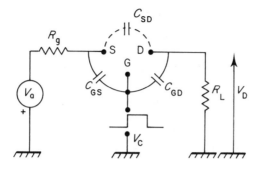

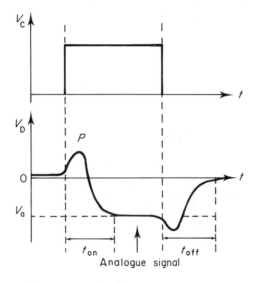

V_c control signal

P transient peak

Figure 6.41   Transient state

be shunted to the ground. The only other component contributing to feed-through would then be C_{ds} and printed circuit board strays.

The MOSFET equivalent circuit in the off state is similar to the JFET except that one must add three more capacitances due to the presence of the body. These consists of C_{sb} and C_{db} in parallel with the generator and load resistance respectively and C_{gb} in parallel with the driver output capacitance.

6.2.5.4 Charge coupling and switching time

The voltage transitions that occur at the gate of a FET switch when it is turned on or off are coupled into the analogue signal path via C_{gs} and C_{gd} and will appear as voltage spikes at the input and output of the switch.

The transient at the input is usually very small if the analogue signal source resistance is low and does not contribute to any significant error to the switch performance. The transient at the output, however, can be detrimental to the system performance.

The magnitude of the voltage spike at the load is dependent on numerous factors:

(1) the amplitude of the driver output voltage transition (V_C);
(2) the driver output voltage transition speed;
(3) the gate to drain and/or gate to source capacitances;
(4) the load impedance;
(5) the analogue generator impedance;
(6) the gate driver impedance.

These factors are complex to determine. For example, FET capacitances are a function of the junction bias voltages whilst FET resistance varies according to the gate-to-channel voltage. Therefore, it is extremely cumbersome to derive the exact expression for the feedthrough transient and thus the switching time.

Since it is apparent that the channel switching time is largely determined by the gate capacitance, a low value of interelectrode capacitance is required for fast switching time. Low capacitance is generally achieved by using smaller-geometry FETs. But this implies that the value of on resistance is likely to increase.

Chapter 7

DIGITAL-TO-ANALOGUE CONVERTERS: ANALOGUE-TO-DIGITAL CONVERTERS

7.1 DIGITAL-TO-ANALOGUE CONVERTERS

The rapid development of new forms of digital-to-analogue converters (DACs) has been made possible with the arrival of integrated analogue switches. The analogue quantity that is the 'output' of a DAC, representing the input digital data, may be a 'gain', also called multiplying DAC, a current, and/or a voltage. In a data-acquisition and -processing system, a DAC may be used as a computer output interface. These devices reconstitute the original data as an analogue quantity after processing, storage, or even simple digital transmission from one location to another.

7.1.2 Principal Types of DACs

7.1.2.1 DAC with weighted resistances

This basic converter consists of:

— An internal reference voltage source E_{ref};
— a set of analogue switches driven by a digital input code stored in a binary register;
— an arrangement of binary-weighted precision resistors that develops a binary-weighted output current.

In order to obtain a substantial voltage output at low impedance, an op amp is required. In this example (Figure 7.1) an op amp holds one end of all the resistors at a virtual signal ground. The switches which are open for 0 or closed for 1, are operated by the digital logic. Each switch that is closed adds a binary-weighted increment of current E_{ref}/R_j via the summing node connected to the amplifier's inverting input. The negative output voltage is

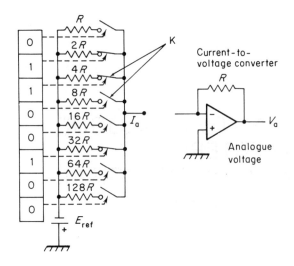

K : Analogue switch

Figure 7.1 Principle of a DAC with weighted resistors

proportional to the total current, and thus to the value of the binary number. The example in Figure 7.1 shows the total output current for an input code 01 100 100 as

$$I_a = E_{\text{ref}}\left(\frac{1}{2R} + \frac{1}{4R} + \frac{1}{32R}\right)$$

Numerical example Suppose an 8 bit DAC with a full-scale output voltage of 10 V is available. Thus, the analogue weight of the LSB (in a series of binary digits, the Least-Significant Bit is the bit that carries the smallest value, or weight) is:

$$1\,\text{LSB} \equiv \frac{10\,\text{V}}{256} = 39\,\text{mV}.$$

Corresponding to the code 01 100 100, we have:

$$V_a = 39 \times 100 = 3900\,\text{mV} = 3.9\,\text{V},$$

because 01 100 100 corresponds to $2^6 + 2^5 + 2^2 = 100$ in decimal code.

Note that the maximum output voltage (10 V − 1 LSB) corresponds to 1111 1111, i.e., $2^8 - 1 = 255$ in decimal code.

In practical applications, where 12-bit digital-to-analogue conversion may be more appropriate, the range of resistance values would be 4096:1, or 40 MΩ for the LSB. If the resistors are to be manufactured in thin-or-thick-film, or integrated-circuit form, such a range would be totally impractical, whilst the cost and size are increased when discrete resistors are used. Consequently these converters are usually limited to 4-bit.

7.1.2.2 DAC with resistance ladders

To reduce the resistance range an array with a limited number of resistors with repeated values may be used. One convenient and very popular form of resistor array which is used is the R–$2R$ ladder. Figure 7.2 shows an example of a 4-bit DAC using an R–$2R$ ladder network with an inverting operational amplifier.

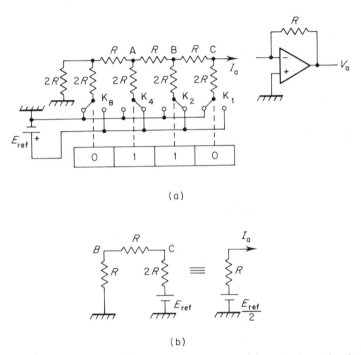

Figure 7.2 (a) DAC with ladder network; (b) equivalent circuit of an element of the ladder network

The switches are operated by the digital input. The resulting output voltage is proportional to the total current flowing towards the amplifier's inverting input.

Assume that i is the current flowing through K1, when all bits but the MSB are off (i.e. grounded). It can be shown that:

— When all bits but bit 2 are off, the current through K2 is $\frac{i}{2}$.
— When all bits but bit 3 are off, the current through K4 is $\frac{i}{4}$.
— When all bits but bit 4 are off, the current through K8 is $\frac{i}{8}$.

Numerical example

$$E_{\text{ref}} = 10\text{ V}, \quad R = 10\text{ k}\Omega.$$

The analogue MSB weight is then

$$i_a = \frac{E_{ref}}{2R} = 0.5 \text{ mA}.$$

For, the Thévenin equivalent of the circuits located to the left of different points A, B, C is R whatever the number of R–$2R$ modules (Figure 7.2b).

Versions of this type of converter may differ from one another in the particular form of their switches. The drawback of this technique is the limitation of speed conversion. Thus when a bit changes, the current flow is inverted in the corresponding $2R$ resistor. Thus, the stray capacitances slow down the conversion.

7.1.2.3 DAC with weighted currents

As the current always flows in the same direction, this type of DAC doesn't present the difficult problem of switching speed, associated with those DACs which use resistance ladders. The weighted current technique consists of the use of binary-weighted current sources, in groups of four, with currents ranging from $8i$ to i. A configuration depicted in Figure 7.3 shows a DAC for BCD coding consisting of three quad current sources with three weighted resistors in series.

The choice of the value of the current i is an important factor. Too large a current increases the power consumption whilst a small current has less noise immunity. A principal advantage of this circuit is its extremely high conversion speed.

DACs with weighted currents are generally designed with bipolar transistors. In Figure 7.4, the binary weighted currents are generated by an R–$2R$ ladder. The reference voltage is applied to the network through transistor T6 and op amp OA1. Amplifier OA2 can be used to convert

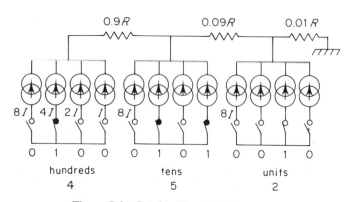

Figure 7.3 DAC with weighted currents

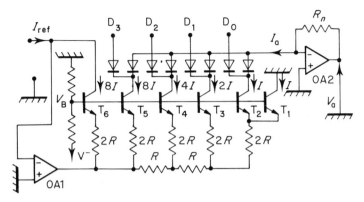

Figure 7.4 Realization of the current sources with bipolar transistors

current i_a into output voltage in accordance with

$$V_a = R_n i_a \quad \text{with} \quad i_a = \frac{i_{\text{ref}}}{8} D \quad 0 \leq D \leq 15.$$

When a positive voltage is applied to the digital input pin, the input diode is forward-biased while the other is back-biased. When a negative voltage is applied, the input diode is reverse-biased while the other is forward-biased, so that the current flow is switched to the summing point.

The accuracy of the binary weighted current source cannot be obtained with discrete components as the V_{BE} drops of the current sources are unequal due to the different collector currents. However IC technology allow the realization of bipolar transistors with weighted emitter areas matched with weighted collector currents as maintaining the same current densities. Therefore, the V_{BE} drops are all equal. Within this family the Motorola MC3408 and the Advanced Micro Devices AM 1408 are some typical monolithic modules. A typical specification for settling time is 250 ns, although loss in settling time will be seen due to the dynamics of an op amp which must be used to provide a low-impedance output signal. In spite of its high conversion speed this type of converter is not very popular due to problem of stability with current sources.

7.1.2.4 DAC with inverted $R-2R$ ladder

This type of converter is currently the most widely used due to its reasonable price and its superior performance. The basic structure still consists of a $R-2R$ network but the binary weighted currents are switched between ground and the op amp summing point. This approach maintains a constant current in each ladder leg independent of the switch state. An

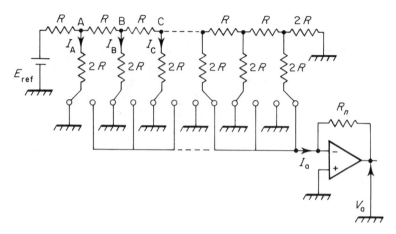

Figure 7.5 DAC with an inverted R–2R ladder structure

important resistor value can be used without altering the conversion speed, thus minimizing errors due to the leakage resistances of the switches.

Referring to Figure 7.5, it can be shown that

$$i_B = \frac{I_A}{2}; \quad i_C = \frac{I_A}{4} \quad \text{and so on} \cdots$$

Especially note that the place of the most significant bit is inverted in comparison with Figure 7.2.

Numerical example

$$E_{ref} = 5 \text{ V} \quad R = 50 \text{ k}\Omega.$$

An R–2R network has the characteristic that the Thévenin equivalent of circuits located to the right of different points A, B, C have output resistance R. Thus the circuit can be redrawn, as shown in Figure 7.6 with the analogue current corresponding to the most significant bit given by

$$i_A = \frac{V_A}{2R} = \frac{E_{ref}}{4R}, \quad \text{therefore} \quad i_A = \frac{5}{200 \times 10^3} = 2.5 \times 10^{-5} \text{ A}.$$

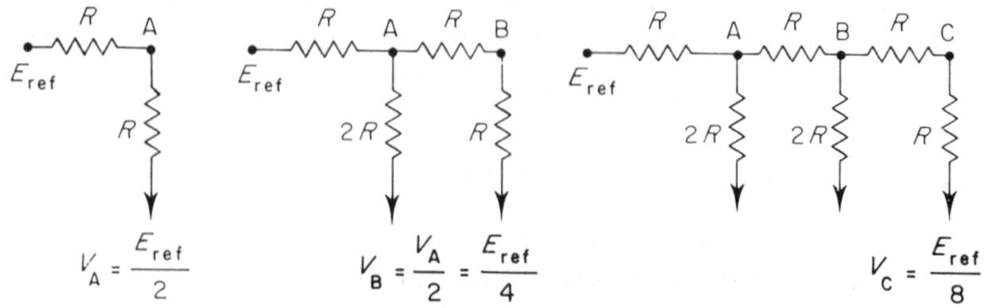

Figure 7.6 Equivalent circuits for the determination of the analog current I_A

As this current is rather small the accuracy of the system is dependent upon the op amp imperfections such as bias current, offset voltage, etc. The Analog Devices AD 7533 is representative of all the characteristics of the other members of this family of converters. A detailed description of a converter of this type follows.

(1) Description of the AD 7533

The AD 7533 is a 10-bit DAC consisting of 10 CMOS switches and a thin-film-on-CMOS R-$2R$ ladder network. The digital input, which is typical of the wide voltage swings of CMOS logic, is also compatible with TTL/DTL logic levels. Besides the 10-bit resolution, the AD 7533 has maximum nonlinearities as low as $\pm 0.05\%$ of V_{ref}, maximum feedthrough error of 1/2 at least-significant bit (LSB = 0.1%) at 100 kHz and typical settling time following a full-scale digital input change of 600 ns.

In addition to a constant or variable reference (current or voltage), of eigher positive or negative polarity, the AD 7533 requires one external op amp for unipolar digitally set gains (2-quadrant multiplication) or two amplifiers for bipolar gains (4-quadrant multiplication).

Figure 7.7 shows a functional diagram of the DAC, which employs an inverted R–$2R$ ladder. Binary-weighted currents flow continuously in the shunt arms of the network. With 10 V applied at the reference input, 0.5 mA flows in the first, 0.25 mA in the second, 0.125 mA in the third, and so on. The $I_{out\ 1}$ and $I_{out\ 2}$ output nodes are maintained at ground potential, either by feedback, or by a direct connection to common. Current is steered to the appropriate output lines by switches which respond to the individually applied logic levels.

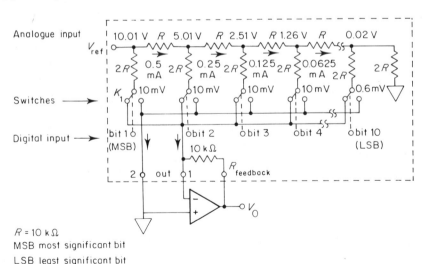

$R = 10\ k\Omega$
MSB most significant bit
LSB least significant bit

Figure 7.7 AD 7533 functional diagram. (Reproduced by permission of D. H. Sheingold, Analog Devices, Inc.)

In the unipolar mode of operation, for a positive reference voltage of 10 V, a 'high' digital input to switch 1 will cause the 0.5 mA of the most-significant bit (MSB) to flow through $i_{\text{out }1}$ and the nominal output voltage of the op amp will be

$$-(0.5\text{ mA}) \times (10\text{ k}\Omega) = -5\text{ V}.$$

With all bits 'on', the nominal output is -9.99 V.
With all bits 'off' the nominal output is zero.

Suppose that all the analogue switches are identical, their resistance value is equal to $R_{\text{ON}}(=100\,\Omega)$. Firstly R_{ON} is not negligible with regard to $2R(=20\text{ k}\Omega)$, and secondly R_{ON} changes with temperature (0.4%/°C). The conversion accuracy is unavoidably affected by the R_{ON} of each switch and their variation with temperature. The R–$2R$ network can also manifest faults which cause the DAC to become nonlinear.

The linearity errors of the AD 7533 are minimized with the following design considerations. The size of each switch is determined so as to arrange an R_{ON} value which is compatible with the binary character of the network. Thus the switch resistance is equal to

20 Ω for the first bit
40 Ω for the second bit
80 Ω for the third bit
...
640 Ω for the sixth bit

This effect provides equal voltages at the ends of the 6 most significant arms of the ladder. As a matter of fact, we can write:

$$10\text{ mV} = 0.5\text{ mA} \times 20\,\Omega = 0.25\text{ mA} \times 40\,\Omega = 0.125\text{ mA} \times 80\,\Omega, \quad \text{etc},$$

As this voltage drop is in series with the reference voltage E_{ref}; it causes an initial 0.1% gain error which does not affect the converter linearity. It is certainly possible to scale the resistance of all the switches to a small enough value which could be negligible with regard to $2R$ in order to eliminate linearity errors and their variation with temperature. However the switches would be physically larger consequently requiring a larger chip.

It is interesting to note that one of the considerations that led to the development of the CMOS technology in a multiplying DAC is that R_{ON} is independent of the polarity of the source to drain current (i_{DS}). This independence is required for use in analogue two-quadrant DACs which can operate with positive and negative currents.

(2) Equivalent Circuits

Figures 7.8 and 7.9 show respectively the equivalent circuit of the AD 7533 at the two following states of input:
all inputs are in 'high state' (Figure 7.8)
all inputs are in 'low state' (Figure 7.9)

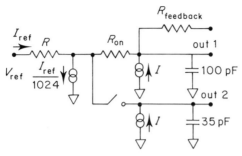

Figure 7.8 AD 7533 equivalent circuit. All digital inputs high

Regardless of the switch states, the input resistance of the ladder network is $R(=10\,\text{k}\Omega)$. The nominal value of I_{ref} is equal to

$$I_{\text{ref}} = \frac{E_{\text{ref}}}{R} = \frac{10\,\text{V}}{10\,\text{k}\Omega} = 1\,\text{mA}.$$

The current source defined by

$$\frac{I_{\text{ref}}}{2^{10}} = \frac{I_{\text{ref}}}{1024} \approx 1\,\mu\text{A}$$

represent one LSB current flowing through the $2R$ terminating resistor (Figure 7.7).

The equivalent resistance of all ten switches connected to:

$I_{\text{out}\,1}$ when all digital inputs are 'high'

and

$I_{\text{out}\,2}$ when all digital inputs are 'low'

is roughly equal to $10\,\Omega(=R_{\text{ON}})$.

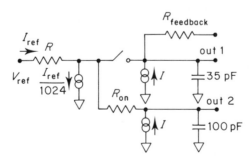

Figure 7.9 AD 7533 equivalent circuit. All digital inputs low. (Reproduced by permission of D. H. Sheingold, Analog Devices, Inc.)

Source current I_{leakage}, which represents junction and surface leakage to the substrate, may be negligible but it changes rapidly with temperature variation.

C_{SD} (source-drain), which is the open-switch capacitance, has a value of about 10 pF.

$C_{\text{out 1}}$ and $C_{\text{out 2}}$ are the output capacitances to ground for the on and off switches.

The equivalent circuits indicate that R_{ON} is distinctly smaller than R_{ladder}. Thus the very small voltage drop accross R_{ON} and the feedthrough coupling via C_{SD} are practically negligible.

There are two most usual forms of AD 7533 applications. The unipolar digital-to-analogue or two-quadrant multiplication version is shown in Figure 7.7, whilst the bipolar offset-binary conversion or four-quadrant multiplication version is shown in Figure 7.11.

(3) Unipolar Conversion

The response equation for Figure 7.7 is nominally

$$V_0 = -\frac{N_{\text{binary}}}{1024} E_{\text{ref}}.$$

Responses to typical codes are tabulated in Figure 7.10. Since E_{ref} may be positive or negative, two-quadrant multiplication is inherent.

(4) Bipolar Conversion

The offset binary response equation for Figure 7.11 is nominally

$$V_0 = -E_{\text{ref}}\left[\frac{N_{\text{binary}}}{512} - 1\right]$$

Responses to typical codes are tabulated in Figure 7.10. If the MSB is complemented, the conversion relationship will be recognized as appropriate for a 2s-complement input, but with a negative scale factor. The MSB determines the sign, and the last 9 bits determine the magnitude in 2s complement notation. Since E_{ref} may be positive or negative, four-quadrant multiplication is inherent.

In this configuration, $I_{\text{out 2}}$, which is the complement of $I_{\text{out 1}}$, is inverted and added to $I_{\text{out 1}}$, halving the resolution (of each polarity) and doubling the gain. The 10.24 MΩ resistor corrects for a 1/1024 difference (inherent in this technique) between $I_{\text{out 1}}$ and $I_{\text{out 2}}$ at zero (1 000 000 000).

The current through the feedback resistor is

$$I_0 = I_{\text{out 1}} - I_{\text{out 2}} \quad \text{with} \quad I_{\text{out 1}} = \frac{E_{\text{ref}}}{R} \times \left[\frac{N_{\text{binary}}}{1024}\right].$$

Besides:

$$I_{\text{out 1}} + I_{\text{out 2}} = \frac{E_{\text{ref}}}{R} - \frac{E_{\text{ref}}}{R}\frac{1}{1024}.$$

Binary code table

Unipolar operation		Bipolar operation	
Digital input	Analogue output	Digital input	Analogue output
MSB LSB 1111111111	$-V_{ref}\left(\dfrac{1023}{1024}\right)$	MSB LSB 1111111111	$-V_{ref}\left(\dfrac{511}{512}\right)$
1000000001	$-V_{ref}\left(\dfrac{513}{1024}\right)$	1000000001	$-V_{ref}\left(\dfrac{1}{512}\right)$
1000000000	$-V_{ref}\left(\dfrac{512}{1024}\right) = -\dfrac{V_{ref}}{2}$	1000000000	0
0111111111	$-V_{ref}\left(\dfrac{511}{1024}\right)$	0111111111	$+V_{ref}\left(\dfrac{1}{512}\right)$
0000000001	$-V_{ref}\left(\dfrac{1}{1024}\right)$	0000000001	$+V_{ref}\left(\dfrac{511}{512}\right)$
0000000000	$-V_{ref}\left(\dfrac{0}{1024}\right)$	0000000000	$+V_{ref}\left(\dfrac{512}{512}\right)$

Figure 7.10 Conversion table

By omitting the term

$$\frac{E_{ref}}{R}\frac{1}{1024},$$

which corresponds to the loss of 1 LSB current through 2R ladder-termination resistor, we have

$$i_{out\,1} - i_{out\,2} = 2\,i_{out\,1} - \frac{E_{ref}}{R}$$

$$= \frac{E_{ref}}{R}\left(2\frac{N_{binary}}{1024} - 1\right).$$

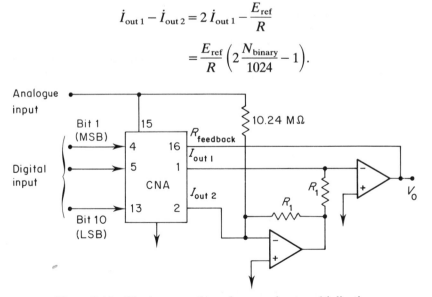

Figure 7.11 Bipolar operation—four-quadrant multiplication

Then:

$$V_0 = -Ri_0 = -E_{ref} \times \left(\frac{N_{binary}}{512} - 1\right).$$

7.1.3 Main DAC Specifications

Some definitions of the performance spectifications are provided in the following list.

7.1.3.1 Resolution

An n-bit binary converter should be able to provide 2^n distinct and different analogue values corresponding to the set of n-bit binary words. A converter that satisfies this criterion is said to have a resolution of n-bits. The smallest output change that can be resolved is 2^{-n} of the full-scale span. Resolution may be expressed as 1 part in 2^n, as a percentage (%), in parts-per-million (p.p.m), or simply by n-bits.

7.1.3.2 Absolute accuracy

The error of a DAC is the difference between the actual analogue output and the output expected when a digital code is applied to the converter. Sources of error include gain error, zero error, linearity error, and noise. It is important to note that error should not be mistaken for resolution.

7.1.3.3 Offset error of a unipolar DAC

This error is the difference between the actual analogue output and the zero analogue output expected when all bits are turned off (binary zero code: 00···000). It may be expressed as a percentage of the full-scale range, or fractions of 1 LSB.

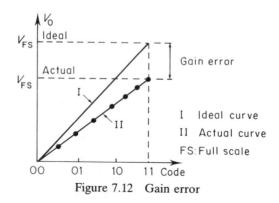

Figure 7.12 Gain error

7.1.3.4 Gain error of a unipolar DAC

Zero-adjustment being made, the gain error is the difference between the actual analogue output and the theoretical full-scale output $E_{ref} \times (1 - 2^{-n})$ expected when all bits are turned on. It may be expressed as a percentage of full-scale range (Figure 7.12).

7.1.3.5 Linearity

The linearity error of a converter, expressed as a percentage—or in p.p.m of full-scale range, is the maximum deviation ε_M of the actual transfer characteristic of the measured conversion relationship, from the best straight line (Figure 7.13).

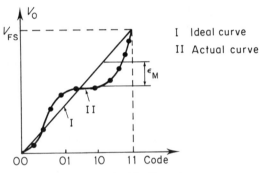

Figure 7.13 Nonlinearity error

7.1.3.6 Differential nonlinearity

Any two adjacent digital codes should produce an output value exactly equal to 1 LSB. Any deviation of the measured 'step' from the ideal difference is called differential nonlinearity, expressed in multiples of 1 LSB. When a differential nonlinearity error is greater than 1 LSB, a non-monotonic response can be produced in a DAC.

7.1.3.7 Monotonicity

The DAC is said to be monotonic when an increase of output voltage results from an increase in input voltage. In the opposite case it is non-monotonic. This error is due to a differential nonlinearity higher than ±1 LSB (Figure 7.14).

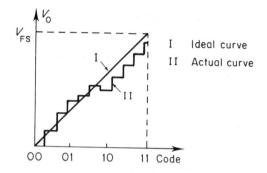

Figure 7.14 Monotonicity error

7.1.3.8 Glitches

When the DAC switches around the MSB, all switches change state, i.e. from 01 111 111 to 10 000 000. For a short time these switch operations cause the DAC to give a large transient spike usually known as *glitch*. In order to remove these glitches, a deglitcher can be used. This consists of a fast sample-and-hold device that holds the output constant until the switches reach equilibrium.

7.1.3.9 Settling time

DAC settling time is a parameter of importance in high-speed applications. Settling time is defined as the time required for the output to approach a final value within the limits of a defined error band for a step change in input. This fixed error band is generally expressed as a fraction of full-scale $\pm \varepsilon \%$ (Figure 7.15). Usually, settling time is the time taken for a DAC to settle after a full-scale code change, to within the analogue equivalent of $\pm \frac{1}{2}$ LSB. It is very important to mention that the settling time of the current-output DAC is quite fast (<300 ns) whereas the settling time of a

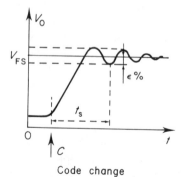

Figure 7.15 Settling time t_s

voltage-output DAC is strongly influenced by the settling time of the op amp circuit. A proper choice of op amp is a primary consideration for fast conversion applications.

7.1.3.10 Conversion rate

The number of conversions per unit time that a DAC may make without violating any of the specifications is called the conversion rate.

7.1.3.11 Influential factors

(1) Temperature coefficients

In general, temperature instabilities are expressed as %/°C, p.p.m/°C, as fraction of 1 LSB/°C, or as a change in a parameter over a specified temperature range. Parameters of interest include gain, linearity, offset (bipolar) and zero.

(2) Long-term Drift

Changes are due to the variations of parameters or specification with time. The most drift sensitive performance parameter is generally the gain error.

7.1.3.12 Notes on high-resolution DACs

The market for high-resolution DACs is growing, due to their widespread presence in measuring systems. These components, which are made more frequently in monolithic technology can now achieve high performance.

An example of a converter of this type is the CMOS 16-bit DAC produced by Matra Harris Semiconductor under the name of HI-DAC 16 B

— power dissipation: 465 mW.
— digital input codes can be either straight binary, offset binary or 2s complement.
— This unit provides output current signals in unipolar operation ranging from 0 to -2 mA.
— Settling time specified to $\pm 0.003\%$ of full-scale range is 1 μs.
— Nonlinearity at 25°C: $\pm 0.0023\%$ of full-scale range.

The high resolution is due to an ingenious disposition of a network of $R-2R$ laser-trimmed resistors on the chip.

Due to the extremely high resolution and linearity of this unit, system design problems such as grounding and contact resistance become very important. For a 16-bit converter with a $+10$ V full-scale range, one LSB is 153 μV. Although the problems involved seem enormous, care in planning its installation can minimize the potential causes of error.

- Analogue output common and power supply common should be tied together at some point in the system as closely as possible to that HI-DAC 16 B to prevent any difference in voltage between them.
- To keep a good settling time, connection between the I_{out} pin and the op amp should be as short as possible to minimize parasitic capacitance.
- For best performance and noise rejection, power supply decoupling capacitors should be located close to the HI-DAC 16 B.
- The choice of the op amp affects the speed and accuracy. Note that another means to improve the DAC resolution exists. Once the system is packaged, errors due to the resistance network may be corrected by a PROM memory.

7.1.4 Examples of Applications

7.1.4.1 Design of a programmable filter

A band-pass filter, which satisfyes the following characteristics, is required

- An adjustable central frequency f_0 between 10 Hz and 10 kHz with a resolution of 10 Hz.
- A 10-bit digital input code determines f_0.
- Constant resonant factor Q equal to 5.

Figure 7.16 shows one of the possible filter circuits. The multiplying module, with factor K in this circuit is an AD 7533 Analog Devices converter used as a two-quadrant multiplying DAC connected as Figure 7.16(b).

The transfer function of the filter circuit is

$$F(p) = \frac{S(p)}{E(p)} = \frac{KR_2\rho_2 C_2 p}{R_1\rho_1\rho_2 C_1 C_2 p^2 - R_2\rho_2 KC_2 p + K^2 R_2}.$$

Therefore

$$Q = \sqrt{\left(\frac{\rho_1 C_1 R_1}{\rho_2 C_2 R_2}\right)} \quad \text{and} \quad \omega_0 = K\sqrt{\left(\frac{R_2}{\rho_1 C_1 \rho_2 C_2 R_1}\right)}$$

When $K \approx 1$, we find approximately $f_0 = 10^4$ Hz.
If we choose $R_1 = R_1 = 10$ kΩ
$C_1 = C_2 = 1$ nF,
then $\rho_1 \approx 80$ KΩ
$\rho_2 \approx 3$ kΩ.

System performances are mainly limited by the offset voltage output, which changes with each digital input code. Sometimes this presents significant values, because the output pin i_{out} sees the Thévenin equivalent

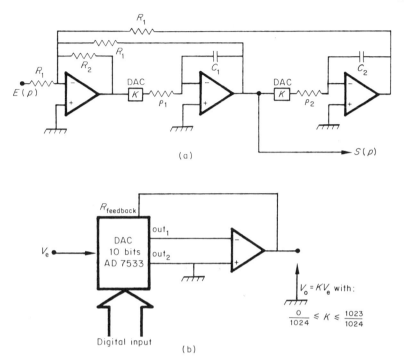

Figure 7.16 (a) Design of a programmable band-pass filter; (b) AD 7533 used as two-quadrant multiplying DAC converter

circuit of the $R-2R$ ladder network as a voltage source in series with an internal resistor whose value depends on the states of the switches.

7.1.4.2 Transistor curves tracer system

We want to visualize the family of transistor collector curves, corresponding to the set of graphs $i_C = f(V_{CE})$ for different values of base current i_B, on an oscilloscope screen.

Increasing values of base current i_B is generated as a step function with N steps. The bias voltage applied to points C and E as shown in Figure 7.18 is generated as a ramp train synchronized with the i_B step function so that V_{CE} increases while i_B remains constant. Both of these signals are generated from an Intel 8085 microprocessor associated with two DACs type AD 7533 as shown in Figure 7.17.

Due to the grounding problem, the signal voltage $R_C i_C$, which is proportional to i_C, must be connected to the oscilloscope via a differential amplifier.

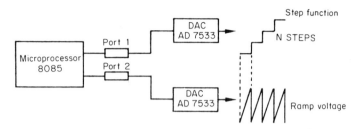

Figure 7.17 Generation of step function and ramp voltage

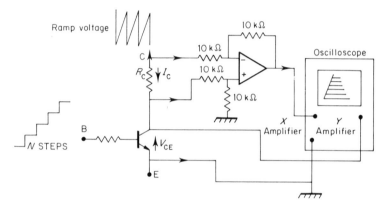

Figure 7.18 The interface between oscilloscope and electronic system is achieved by a differential amplifier

7.2 ANALOGUE-TO-DIGITAL CONVERTERS

In this chapter, we suppose that all the measuring variables have been converted into voltages. Then, analogue-to-digital converters (ADCs) convert these analogue input voltages into its equivalent digital form.

There are a vast number of conceivable circuit designs for ADCs. A much more limited number of designs which are specially designed for incorporation as components of equipment are available on the market in small, modular form at low cost.

The most popular of these are:

— parallel types;
— successive-approximation types;
— pulse-counting types.

7.2.1 Parallel ADCs (Flash Converters)

The technique of this converter type consists of comparing the analogue input voltage E_X with n reference voltages simultaneously. Figure 7.19

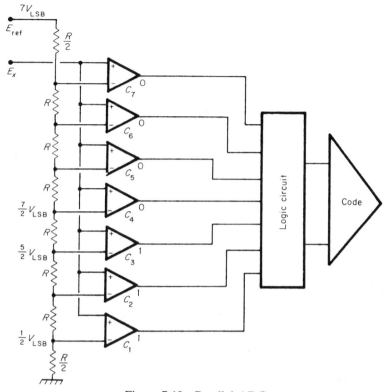

Figure 7.19 Parallel ADC

shows an example of a parallel 3-bit converter. Eight different binary output numbers are possible using seven comparators and seven reference voltages generated by resistances voltage-dividers. For 0 input, all comparators are off. As the input increases, it causes an increasing number of comparators to switch state. The smallest voltage that can switch on the first comparator is

$$\tfrac{1}{2}V_{\text{LSB}} = \tfrac{1}{2}\frac{E_{\text{ref}}}{7}.$$

Take the example of an input voltage E_X within the range

$$\tfrac{5}{2}V_{\text{LSB}} < E_X < \tfrac{7}{2}V_{\text{LSB}}$$

Comparators 1–3 are on and the others are off. The outputs of these comparators are then applied to the gates, which provide a set of outputs that fulfill the approximate condition for natural binary output. Other codes such as the Gray code are also possible.

In this example, the number 3(011) must be displayed. The evident advantage of this approach is that conversion occurs in parallel, with speed limited only by the switching time of the comparators and gates. With ECL

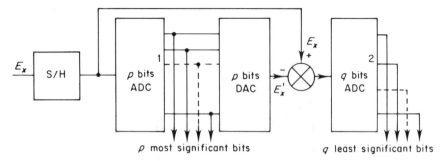

Figure 7.20   Use of two parallel ADCs for improving resolution

logic circuits, the control clock can reach 100 MHz. Thus, this is the fastest approach to conversion. Unfortunately, the number of elements increases geometrically with resolution. For a n-bit converter, $2^n - 1$ comparators are needed. Therefore, high resolution and the fastest speeds at low cost are still some time away.

When the speed of conversion is not essential, the circuit of a modified-parallel ADC depicted in Figure 7.20 shows that the number of comparators can be reduced while the resolution is improved.

In this circuit scheme, a p-bit ADC gives the first p most significant bits. This portion of converted voltage is reproduced by a DAC, and then it is subtracted from the initial input voltage E_X. A second q-bit ADC is used to convert the difference $(E_X - E'_X)$, thus giving the q least significant bits.

The whole system is equivalent to an n-bits ADC, where $n = p + q$. As E_X and E'_X should be related to the same input signal, an external sample-and-hold is required to keep the input signal unchanged throughout the conversion.

Flash converters are coming into widespread use, since they provide the promise of instantaneous conversion.

Examples of industrial product
Model CAV-1210 (Analog Devices)
Features:

— 12-bit resolution
— 225 ns conversion time
— 10 MHz conversion rate.

Figure 7.21 shows two conversion steps. Basically, the correction circuits use the information contained in the MSB of the second conversion step result to determine what action needs to be taken with regard to the first six bits. Depending upon its value, the circuits will pass the 6-bit information as it is, or add a value of binary 1 to it. Bits 2–7 of the 7-bit information become Bits 7–12 of the digital output. The use of 13-bits to achieve an accurate 12-bits of resolution allows compensation for a multitude of errors which would be eliminated by incorporating high-precision parts into the design, but this might be very complex and expensive.

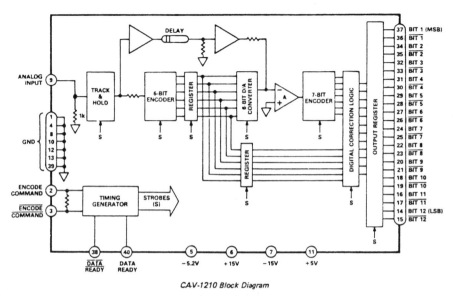

CAV-1210 Block Diagram

Figure 7.21 CAV-1210 block diagram. (Reproduced by permission of D. H. Sheingold, Analog Devices, Inc.)

In the general case, it is advisable to use a track/hold ahead of the converter.

Effectively, a typical comparator cell consists of a differential input with *analogue input signal* on one side and a *reference voltage* on the other. Certain capacitance characteristics of these cells tend to degrade the performance of the converter when the analogue input frequency is increasing. The analogue input capacitance C_{bei} of each cell is determined by the base–emitter junction capacitance. The value of this capacitance is affected by junction bias with forward-bias capacitance higher than reverse-bias capacitance. Since the inputs of all capacitors are in parallel, the total capacitance at the analogue input is the sum of the individual comparators base capacitances, this sum varies from 30 to 120 pF because it depends on the amplitude of the analogue signal. The reference input is similarly subject to capacitance variations introduced by its base–emitter capacitance C_{ber}. The charging path for each individual C_{ber} is through the reference ladder.

The time required to charge the capacitances through the resistors tends to cause the reference voltage to lag fast changes in the current flowing through the reference ladder. Then the use of a track/hold ahead of the flash converter is advisable. It is important to remember that a track/hold amplifier has no effect on the conversion time. Its function is to follow all changes in the analogue input as they occur ('track' mode) and to freeze ('hold' mode) an instantaneous sample of that input while the comparators

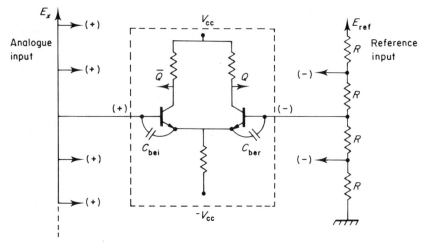

Figure 7.22 Typical comparator cell

are latching. It can also be timed to allow the comparator input capacitances to charge and settle before the conversion takes place.

7.2.2 Successive-approximation ADCs

7.2.2.1 Functional principle

Successive-approximation ADCs (Figure 7.23) are quite widely used, especially for interfacing with computers, because they are capable of both high resolution (up to 16 bits) and high speed (up to 1 MHz throughput rates). Conversion time, being independent of the magnitude of the input voltage, is fixed by the number of bits in the register and the clock rates. Each conversion is unique and independent of the results of previous conversion, because the internal logic is cleared at the start of a conversion.

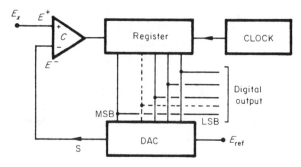

Figure 7.23 A/D converter using the successive approximations technique

Modern *IC* converters include three-state data outputs and byte controls to facilitate interfacing with microprocessors. A three-stage output has, in addition to the normal 1 and 0 state, a not-enabled condition, in which the output is simply disconnected via an open voltage switch. This permits many device outputs to be connected to the same bus, only the device that is enabled (one at a time) being able to drive the bus. Since typical processor data buses are only 8 bits wide, 10 or 12-bit data must be communicated in two steps, one byte at a time.

The conversion technique consists of comparing the unknown input against a precisely generated internal voltage at the output of a DAC. The input of the DAC is the digital number at the output of the ADC. The conversion process is strikingly similar to a weighing process using a chemist's balance, with a set of n binary weights (e.g., $\frac{1}{2}$ lb, $\frac{1}{4}$ lb, $\frac{1}{8}$ lb, ... for unknowns up to 1 lb).

After the conversion command is applied, and the converter has been cleared, the DAC's MSB output (half full scale) is compared with the input. If the input is greater than the MSB, it remains on (i.e., 1 in the output register), and the next bit (quarter full scale) is tried. If the input is less than the MSB, it is turned off (i.e., 0 in the output register), and the next bit is tried. The process continues in order of descending bit weight until the last bit has been tried. When the process is completed, the status line changes state to indicate that the contents of the output register now constitute a valid conversion. The contents of the output register form a binary digital code corresponding to the input signal.

Saying that the binary output number $N_0 = B_1 B_2 B_3 \cdots B_n$ is related to the analogue input voltage E_X means

— in unipolar operations

$$E_X = E_{\text{ref}} \frac{B_1 2^{n-1} + B_2 2^{n-2} + \cdots + B_n 2^0}{2^n} \qquad (7.1)$$

— in bipolar operations: as a fixed bipolar offset equal to $E_{\text{ref}}/2$ is internally summed with E_X so that the total will be positive over the rated operating range, expression (7.1) becomes

$$E_X = E_{\text{ref}} \frac{B_1 2^{n-1} + \cdots + B_n 2^0}{2^n} - \frac{E_{\text{ref}}}{2}.$$

An example will illustrate how the binary coding is made (Figure 7.24). Suppose that we have a 3-bit DAC used for unipolar input.

Phase 1. The input voltage E_X is compared with the MSB weight voltage $E_{\text{ref}}/2$:

If $E_X > \dfrac{E_{\text{ref}}}{2}$, then $B_1 = 1$, otherwise $B_1 = 0$.

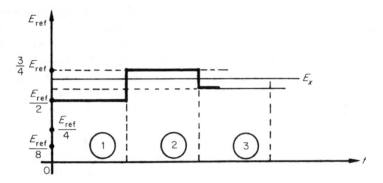

Figure 7.24 Timing Diagram of a 3 bit conversion

Phase 2: E_X is then compared to $B_1 E_{ref}/2 + E_{ref}/4$:

$$E_X < \frac{E_{ref}}{2} + \frac{E_{ref}}{4}, \quad \text{then} \quad B_2 = 0.$$

Phase 3:   Next, E_X is compared to $B_1 E_{ref}/2 + B_2 E_{ref}/4 + E_{ref}/8$:

$$E_X > \frac{E_{ref}}{2} + 0 + \frac{E_{ref}}{8}, \quad \text{then} \quad B_3 = 1.$$

The process completed, the final result is 101, i.e.

$$E_X = E_{ref} \times \frac{(101) \text{ binary}}{2^3}$$

The difference between the actual input voltage E_X and the voltage corresponding to the binary output code is not cancelled but only limited to less than 1 LSB.

The successive-approximation converter has the weakness that at higher rates of change, it generates substantial linearity errors because it cannot tolerate change during the weighting process. In this case, the converted value lies somewhere between its values at the beginning and the end of conversion. That is why it is usual to employ a sample-and-hold device ahead of the converter to retain the input value that was present at a given time before the conversion starts, and maintain it constant throughout the conversion. The status output of the converter could be used to release the sample-and-hold from its hold mode at the end of conversion. Even if the signal is slow enough, noise rates of change that are excessively large will cause erroneous readings that cannot be averaged by either analogue or digital means. Thus an external sample-and-hold may still be needed.

An association of an ADC with a sample-and-hold amplifier for high-speed signal acquisition is shown in Figure 7.25.

The normal stand-by situation is shown at the left-hand end of the

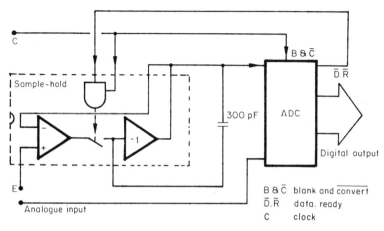

Figure 7.25 Sample-Hold device connections to an ADC

drawing in Figure 7.26. The Blank and $\overline{\text{Convert}}$ (B & $\overline{\text{C}}$) line is held high, the output lines will be 'open', and the $\overline{\text{Data Ready}}$ ($\overline{\text{DR}}$) line will be high. This mode is the state consuming the lowest power of the device. When the B & $\overline{\text{C}}$ line is brought low, the conversion cycle is initiated. The $\overline{\text{DR}}$ and Data lines do not change state until the conversion cycle is complete, when the $\overline{\text{DR}}$ line goes low and the data lines become active with the new data.

About 1.5 μs after the B & $\overline{\text{C}}$ line is again brought high, the $\overline{\text{DR}}$ line will go high and the data lines will go open. The minimum pulse width for the B & $\overline{\text{C}}$ line to blank previous data and start a new conversion is about 2 μs.

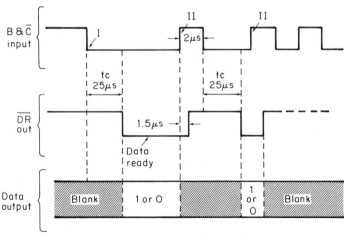

I : Start conversion   II : Stop conversion

tc : Conversion time

Figure 7.26 Example of control and timing of an ADC

Example of industrial devices
AD 571: 10-bit ADC AD 571 (Analog Devices)
  25 μs conversion time
  3-state output
  18-pin ceramic dip
AD 582: Sample-and-hold amplifier
  Suitable for 10-bit ADC
  Low acquisition time: 6 μs to 0.1%
  Low charge transfer: <2 pC.
  The AD 582 acquire a 10 V signal in less than 10 μs with a droop rate less than 100 μV/ms. It is driven to the 'hold' mode when the (B & C̄) line goes low. The Data Ready line is fed back to the other side of the differential input control gate so that the AD 582 cannot come out of the 'hold' mode during the conversion cycle. At the end of the conversion cycle, the Data Ready line goes low, automatically placing the AD 582 back into the sample mode.

7.2.2.2 Examples of application in instrumentation

Analog Delay Line

The problem is to conceive a pure time delay device for an analogue signal. The input signal will be a sine wave with a positive peak of 5 V and a variable frequency varying from 0 to 500 Hz. The block diagram of the circuit is represented in Figure 7.27. The delay is produced by m-bit shift registers for n parallel digital channels, each channel representing 1-bit of converted analogue signal. The right shifting occurs at the positive front of the clock pulse. The delay time is controlled by the clock frequency f_s; it is

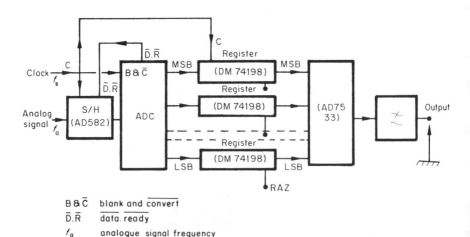

Figure 7.27 Precision analogue delay line

determined by the following equation:

$$\text{Delay time} = m\frac{1}{f_s} + \text{conversions time.}$$

7.2.2.3 Microprocessor-compatible ADC

The organization of the simplest microprocessor compatible ADC is shown in Figure 7.28. It includes:

— Five control inputs: chip-select ($\overline{CS}$)
　　　　　　　　　　　chip-enable ($\overline{CE}$)
　　　　　　　　　　　read/write (R/$\overline{W}$)
　　　　　　　　　　　format (binary or 2s complement coding)
　　　　　　　　　　　bipolar/unipolar analogue range (BPO/$\overline{UPO}$)
— a status output which indicates when conversions are in progress.

The microprocessor can initiate conversions by bringing R/$\overline{W}$, $\overline{CS}$, and $\overline{CE}$ all low. At the beginning of a conversion, the states of the FORMAT and BPO/$\overline{UPO}$ input lines are latched into the converter. The FORMAT and BPO/$\overline{UPO}$ inputs can be either hard wired to the appropriate logic level or software-programmed for each conversion.

When a conversion is initiated by the microprocessor, the Status line goes high and returns low only at the end of the conversion. Then the microprocessor can read and place the results of the conversion on the data bus by bringing $\overline{CS}$ and $\overline{CE}$ low, while R/$\overline{W}$ is high. The timing diagrams for *Convert* and *Read* cycles are shown in Figure 7.29.

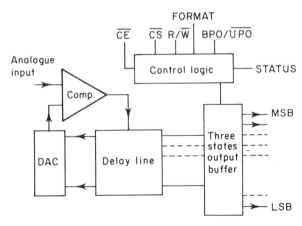

Figure 7.28 Organization of a standard microprocessor-compatible ADC

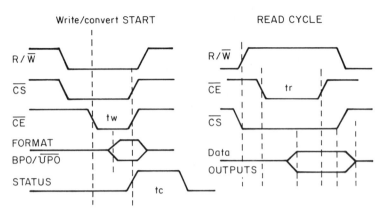

Figure 7.29 Timing diagrams for convert and read

Example of industrial product
Model AD 670 (Analog Devices)
Features: 8-bit resolution and accuracy
    10 μs conversion time
    single +5 volts supply
    flexible input stage: Instrumentation amplifier front end provides differential input and good CMR.

7.2.3 Impulse-counting Converters

Analogue-to-digital conversion by the counter method requires the least components giving a high accuracy with relatively simple circuits. However, conversion time, which is considerably longer than with the other methods, is usually between 0.1 and 100 ms. This is quite sufficient for the many applications in which it is the most widely used type. The most important circuits from the extensive variety are,

— single-slope converters;
— voltage-to-frequency converters;
— dual-slope converters;
— three fold-slope converters;
— quad-slope converters.

7.2.3.1 Single-slope converters

The operating principle of the single-slope ADC is based on the measurement of the time it takes for a constant slope linear ramp voltage to rise from 0 V to the level of the input voltage E_x, or to decrease from the level of the input voltage to zero. This time interval, which is measured by counting the pulses of a quartz oscillator, is proportional to the input voltage.

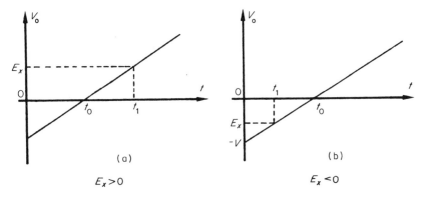

Figure 7.30 Comparison between the input voltage and a linear ramp voltage: (a) $E_x > 0$; (b) $E_x < 0$

(1) Example of a positive input converter

The conversion is illustrated by the Figure 7.30. At the start of the measurement cycle (time = t_0), a positive ramp voltage is initiated from 0 V. At the same time, the gate is opened, thus allowing clock pulses from an oscillator to be counted by a number of decade counting units which totalize the number of pulses passed through the gate.

The positive-going ramp, shown in Figure 7.31, is continuously compared with the unknown input voltage. At the instant t_1 that the ramp voltage equals the unknown voltage, the comparator generates a pulse which closes the gate.

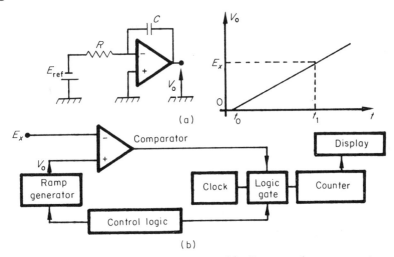

Figure 7.31 Single slope converter: (a) Ramp voltage generator $V_0 = \dfrac{E_{\text{ref}}}{RC}(t_0 - t)$; (b) functional diagram

The number of counted clock pulses during the time interval $(t_1 - t_0)$ is directly proportional to the unknown input voltage E_x as:

$$E_x = \frac{E_{ref}}{RC}(t_1 - t_0) \quad \text{or} \quad E_x = \frac{E_{ref}}{RC} NT$$

where T is the clock period, N is the number of counted pulses.

(2) Example of a bipolar input converter

The basic circuit is shown in Figure 7.32. The system includes two comparators A, B and a sawtooth generator. The sawtooth voltage rises from negative to positive values according to the expression

$$V_0 = \frac{E_{ref}}{RC} t + V(0).$$

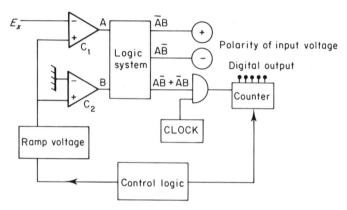

Figure 7.32 Functional diagram of a bipolar input converter

The output of the logic circuit, which performs the Ex-OR operation, $\bar{A}B + A\bar{B}$, is at logic 1 for as long as the sawtooth voltage is between the limits 0 and E_X. The corresponding time interval is therefore given as

$$\Delta t = \frac{RC}{E_{ref}} E_X.$$

It is measured by counting the pulse of a quartz oscillator. If the counter is reset to zero at the beginning of each conversion cycle, the counter state is at

$$N = \frac{\Delta t}{T} = \frac{RC}{E_{ref}} E_X f \quad \text{with} \quad f = \frac{1}{T} \tag{7.2}$$

when the voltage exceeds the upper trigger level of the comparator circuit.

For a negative input voltage, the sawtooth voltage first crosses the input voltage and then passes through zero. Obviously, the sequence is reversed for a positive input voltage.

Therefore, a logic circuit can be used to determine the polarity of the

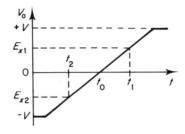

Different output states of comparators				
		$t < t_0$	$t_0 < t < t_1$	$t > t_1$
$E_{x1} > 0$	A	0	0	1
	B	0	1	1
	$\bar{A}B$	0	1	0
		$t < t_2$	$t_2 < t < t_0$	$t > t_0$
$E_{x2} < 0$	A	0	1	1
	B	0	0	1
	$A\bar{B}$	0	1	0

Figure 7.33 Bipolar input converter: determination of input voltage polarity and gate operation

input voltage and the gate operation. Its output will be $\bar{A}B$, $A\bar{B}$ and $\bar{A}B + A\bar{B}$ according to the truth tables established below (Figure 7.33).

Note that the conversion time depends only on the magnitude of the input signal. As is apparent from Eq. (7.2), the accuracy of the conversion result is subject to the temperature drift and long-term stability of the reference generator E_{ref}, the sawtooth generator and the quartz oscillator. For this reason, an accuracy of better than 0.1% is very difficult to attain.

Filters can be used on the input side of the ADC to remove unwanted components of the input signal. Noise and line-frequency pick up are also attenuated in this way, but at the expense of reduced response to fast input-signal amplitude variations. A more convenient way is to average noise by applying the input signal to an integrator as in the voltage-to-frequency method described below.

7.2.3.2 Voltage-to-frequency converters (VFCs)

(1) Functional Principle
Voltage-to-frequency conversion is a simple and low-cost process. The analogue voltage level is converted into pulse trains, square wave or sawtooth waveform in a logic-compatible form (TTL/DTL) at a repetition rate that is accurately proportional to the amplitude of the input signal.

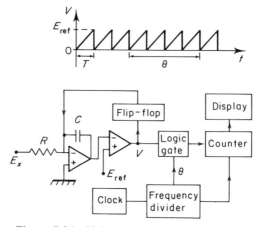

Figure 7.34 Voltage-to-frequency converter

The most popular VFC designs (Figure 7.34) contain an integrator which charges at a rate proportional to the value of the input signal E_X. Each time the integrator's charge reaches a certain potential E_{ref}, a comparator drives the integrator to reset mode which discharges the integrating capacitor. The next pulse is triggered when the net integral has again reached the threshold. Since the time required to reach the switching threshold is inversely proportional to the analogue input, the frequency is therefore directly proportional to it.

$$E_{ref} = \frac{1}{RC} E_X T;$$

hence

$$E_X = RCE_{ref} f, \qquad (7.3)$$

where pulse rate f is determined by counting the numbers of pulses N during a fixed time interval θ. As $f = N/\theta$ equation (7.3) becomes:

$$E_X = \frac{RC}{\theta} E_{ref} N.$$

The major advantage of this method is that with a fixed averaging period θ, high frequencies that have whole numbers of cycles during the averaging period will be nulled out. Therefore, *'infinite' normal-mode rejection* at frequencies that are integral multiples of $1/\theta$ can be obtained. For example, if one wishes to reject line-frequency 50 Hz and its harmonics, integration time θ must be 20.00 ms or a multiple.

However, converter accuracy depends on the precision and the long-term stability of the integrator, the reference voltage and the quartz oscillator. The dual-slope method eliminates most of these inconveniences.

Examples of industrial realization We can find in various manufacturers' general catalogues a set of voltage-to-frequency converters whose maximum output frequencies ranging from 10 KHz to MHz. Some specifications of certain Analog Devices VFCs are shown in Table 7.1

Table 7.1

Maximum full-scale frequency	Model	Nonlinearity error	Stability
10 KHZ	450	0.005%	25 ppm/°C
100 KHZ	AD 537	0.07%	50 ppm/°C
1 MHZ	AD 650	0.01%	100 ppm/°C

(2) Description of the AD 537

The AD 537 is the most used voltage-to-frequency converter because it is both low in cost and easy to use. It is a complete VFC requiring only an external RC timing network to set the desired full-scale frequency and a selectable pull-up resistor for the open-collector output stage. The full-scale frequency is set by the timing capacitor and resistor from the simple relationship.

$$f = \frac{E_X}{10\,RC}.$$

A block diagram of the AD 537 is shown in Figure 7.35. An op amp (BUF) serves as the input stage. When the input voltage is applied at pin 5, the input op amp provides a very high input impedance (250 MΩ).

The input voltage can be converted to the proper current in the transistor follower (at pin 3) by choosing an adequate scaling resistor.

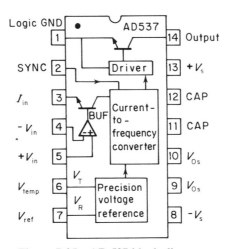

Figure 7.35 AD 537 block diagram

As the input amplifier presents very high input impedance, it allows direct signals coming from most of the usual transducers such as thermocouples, strain gauges, thermistors, photosensors, etc.

The drive current is delivered to the current-to-frequency converter, which is the heart of the VFC device. It provides both the bias levels and the charging current to the externally connected timing capacitor CAP. The oscillator provides a square wave.

The excellent temperature characteristics and long-term stability of the AD 537 are guaranteed by the precision voltage reference which provides a compensation voltage proportional to the temperature drift. The generator also provides a constant reference of 1.00 V output which is very useful to drive resistive sensors and a temperature-proportional output scaled to 1.00mV/K which enables the circuit to be used as a reliable temperature-to-frequency converter.

The output stage is designed for interfacing to all digital logic families (TTL, ECL, CMOS, ...). As the collector and emitter of the output transistor are both uncommitted the emitter can be tied to any voltage between $-V_S$ and 5 V below $+V_S$. The open collector can be pulled up to a voltage 36 V above the emitter regardless of $+V_S$. The output transistor can supply up to 20 mA which is a large enough current to drive up to 12 TTL loads or LEDs in optical isolation application. The logic ground pin can be connected to any level between ground (or $-V_S$) and 4 V below $+V_S$. This allows very easy interfacing to any logic with either positive or negative logic levels.

Unlike most VFCs, the AD 537 can easily be connected in a phase-lock loop which allows considerable extension of the application field of the device.

(3) Interfacing the AD 537

Voltage-to-frequency converters with nonlinearity error less than 0.01% can be used in applications where 12-bit ADCs are desired. It allows analogue-to-digital conversion of small analogue signals coming from transducers without any other signal conditioning. A VFC followed by a binary counter is a real ADC.

If the data must be transmitted through a noisy environment, and over distances greater than few metres then parallel data transmission becomes expensive because of the high cost of wire and wiring, and the need of more driver power to deal with increased line capacitance. Immediately conversion to frequency information, which permits serial transmission, is highly desirable. The VFC pulses require a single wire pair for transmission, unlike parallel converters, which for n-bits, require at least $n + 1$ wires. VFCs may be designed to work with a large number of different transducers. Some examples can illustrate its versatility:

(a) Figure 7.36 shows a thermocouple with a reference junction maintained at constant temperature (T_{ref}) connected to a AD 537 converter.

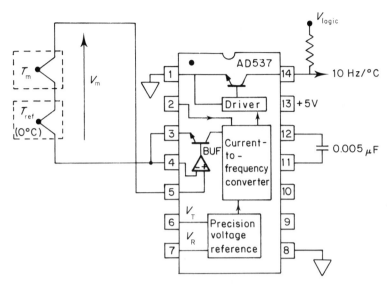

Figure 7.36 Temperature-to-frequency conversion

Full-scale frequency output is perfectly adjustable for a maximum voltage, corresponding to a desired temperature range.

(b) Figure 7.37 shows a strain-gauge bridge driven by the constant 1.00 V reference output via the unity gain operation amplifier which is used to isolate the AD 537 reference output from the bridge.

(c) Most types of resistive transducers can be directly driven from the AD 537 reference output. Figure 7.38 shows a displacement transducer. The

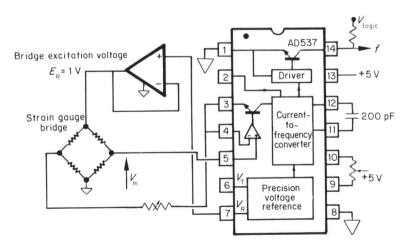

Figure 7.37 AD 537 converter associated with a strain gauge bridge

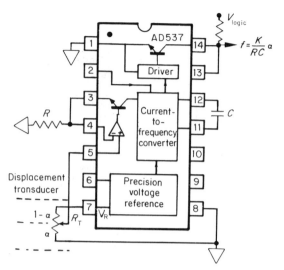

Figure 7.38 Displacement-to-frequency conversion. (Reproduced by permission of D. H. Sheingold, Analog Devices, Inc.)

sense point of the potentiometer is connected to the non-inverting input of the buffer amplifier. The frequency output is given by the expression

$$f = \frac{K}{CR}\alpha$$

where K is the scale correction factor. Its variation which is a function of the total resistance values is shown as a function of R_T in Figure 7.39. Note that by using a panel potentiometer for R_T, the circuit scheme of Figure 7.39

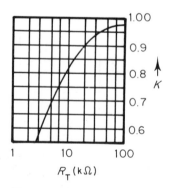

Figure 7.39 The scale correction factor is a function of R_T

provides a very linear variable oscillator without needing adjustable timing components.

(d) By combining the VFC with a floating power supply and optical isolator, accurate low level measurements in the presence of high common-mode voltage may be achieved. Cost and complexity are minimized since only a single isolator is required to couple the serial pulse output from the VFC to the digital read-out. For long-distance data transmission, optical fibre can be conveniently used. This technique is becoming commonly employed in noisy industrial environments.

7.2.3.3 Dual-slope converters

(1) Functional Principle

Another conversion method, in which not only the reference voltage but also the input voltage is integrated, is illustrated by the block diagram in Figure 7.40. The dual-ramp type, which is especially suitable for use in digital voltmeters, operates in the following manner.

Phase 1. At the beginning of the conversion, the input signal is applied to an integrator which integrates for a constant time interval t_1. At the end of the integration interval t_1, the output voltage of the integrator is given by

$$-\frac{1}{RC}\int_0^{t_1} E_X \, dt = -\frac{E_X}{RC} N_1 T, \qquad (7.4)$$

where N_1 is the number of clock predetermined pulses (let's say 1000) and T the period of the clock generator.

Phase 2: The reference voltage is then applied to the integrator. The polarity of the reference voltage is opposite to that of the input voltage so that the magnitude of the integrator output voltage decreases, as can be seen in Figure 7.40. At the same time, the counter is again counting from zero. The comparator and the counter are now used to determine the time interval required for the integrator voltage to reach zero again.

$$\frac{E_X}{RC} N_1 T = \frac{E_{\text{ref}}}{RC} NT.$$

Hence the end result is given by

$$E_X = E_{\text{ref}} \frac{N}{N_1}.$$

Dual-slope integration has many advantages. Conversion accuracy is independent of both the capacitor value and the clock frequency, because they affect both the up-slope and the down-ramp in the same ratio. The only condition is that the clock frequency remains constant during the interval t_2. This short-term stability presents no problem, even for simple clock generators. As is shown in Eq. (7.4), the result is influenced not by the

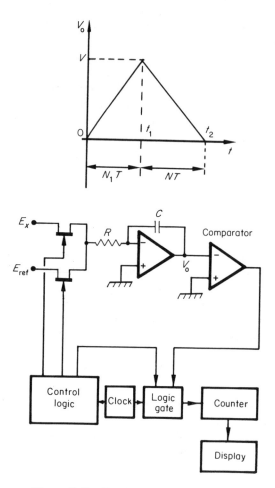

Figure 7.40 Dual slope integration ADC

instantaneous input voltage but by its mean value taken over the measuring time t_1. Alternating voltages are therefore either attenuated or completely rejected if their frequencies are multiples of $1/t_1$. For these reasons, the method can produce an accuracy of better than 10^{-4}, a resolution of $1\,\mu\text{V}$, a nonlinearity of 0.005% FS, an input impedance about $G\Omega$, a common-mode rejection of 150 dB for d.c. input and 100 dB for a.c. input.

Although too slow for fast data acquisition, dual-slope converters are quite adequate for very accurate measurement using industrial transducers. They are also the predominant circuit used in constructing digital voltmeters.

(2) Performance and Limitations

It has been shown for the dual-slope method that the time constant RC and the clock frequency $f = 1/T$ do not influence the result. The accuracy is therefore defined essentially by the tolerance of the reference voltage and the actual performances of integrator and comparator.

(a) *Reference voltage source stability.* The reference voltage is generated by a regulator using a Zener diode. Good regulation requires the Zener diode voltage to be about 6.8 V, so that the Zener resistance has the smallest value. The Zener temperature coefficient varies from -0.062% °C to $+0.06\%$ °C, thus passing through zero% °C for Zener voltage values limited between 5.1 V and 5.6 V. In practical applications, the voltage, which is most suited for the regulation, is selected. The temperature coefficient can be reduced by series connecting forward-biased diodes. Such temperature compensated reference elements have temperature coefficient less than a few ppm/°C.

(b) *Switch imperfections.* When turned on, JFET switches present a small but nonzero value of drain-source resistance (typically 10 to 200 Ω). When the switch is turned off, small inverse leakage currents flow from drain to gate and from source to drain. These currents can be neglected. However, R_{ON} resistance must be taken into account since it adds up to the integrator's resistor R. Of course, this influence can be minimized by increasing the R value. As this reduces the amplifier input current, the op amp bias current is then noticeable. As the R_{ON} value changes with the temperature ($+0.7\%$ °C), the device fidelity is compromised.

(c) *Integrator imperfections.* Two major error sources are:
— input offset voltage;
— nonlinearity of the integrating ramp.

If the input offset voltage V_{iO} is taken into account in Eq. (7.4) then its exact version is given as

$$(E_X + V_{iO}) = \frac{(E_{ref} - V_{iO})}{N_1} N. \qquad (7.5)$$

Note that V_{iO} affects E_X and E_{ref} differently because E_X and E_{ref} have

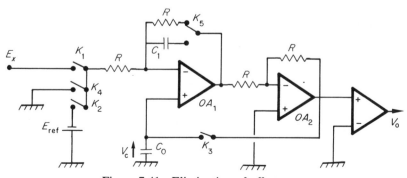

Figure 7.41 Elimination of offset errors

different polarities. This offset voltage changes with the temperature with variations reaching hundreds of $\mu V/°C$. As the sensitivity of a voltmeter with a reference of 5 V and a resolution of 20.000 points is given as

$$\frac{5}{20.000} = 250 \ \mu V.$$

it can be seen that a temperature change of a few °C can cause a change in the least significant bit. An automatic zero, as incorporated in most high-quality volmeters, can eliminate these offset errors. At rest, switch K3 is closed and switch K5 on the resistance R side. The two inverters OA1 and OA2 constitute a voltage follower, whose output voltage V_C charges the zeroing capacitor C_0. For offset correction, the integrator input is simultaneously grounded via switch K4. The voltage V_C therefore assumes the correction value $V_{iO} - \dot{I}_B R$, where V_{iO} is the integrator offset voltage and $\dot{I}_B$ the integrator input bias current.

For the integration of the input voltage, switches K3 and K4 are opened and K1 closed. As the voltage V_C remains stored on the capacitor C_0, the offset is balanced during the integration interval. The zero drift is then only influenced by the short-term stability.

The ramp nonlinearity is essentially due to the leakage resistance of the capacitor C_1 and to the value of amplifier gain. An integrator with nonlinearity of 0.001% is quite feasible today.

(d) *Comparator imperfections.* The ideal comparator would have infinite gain and zero comparison time. Practical comparators are limited by parameters similar to those of op amps such as open-loop gain, bandwidth, input offset voltage and bias current. Normally the amplifier output impedance is sufficiently small so that the input bias current effect is negligible. If A is the finite-gain value, the output voltage V_0 is attained for values of input voltage greater than or equal to

$$V_0 = -\frac{S}{A}.$$

Therefore, an error on the comparison level occurs. Offset error and finite gain both influence the comparator accuracy. Besides gain, d.c. offset and drift, another specification that must be considered in the selection of a comparator is its speed. Unfortunately, speed and precision cannot always be obtained simultaneously. A good compromise nowadays is a few mV for offset voltage and hundreds of nanoseconds for response time.

Examples of industrial realizations Most ADCs of this type are available in CMOS technology. One must differentiate between two major groups:

— Those with the straight-binary coded parallel outputs such as 13-bits AD 7550 (Analog Devices) 8-bits $\cdots$ 12-bits ADC-EK 8B $\cdots$ 12B (Datel)
— Those with BCD outputs: $3\frac{1}{2}$ digits ADC 1100 (Analog Devices) $3\frac{1}{2}$ digits MC 14 433 (Motorola)

7.2.3.4 Threefold slope converter

The first step of the dual-ramp method is included. The next step consists of two more phases.

First phase

A rapid capacitor discharge begins at instant t_1. This stops when the integrator output voltage reaches a defined threshold voltage with the value $V_{\text{threshold}}$ (=100 mV for example). This accelerated discharge is obtained by replacing the integrator input resistance R by a smaller one R'. For example, if R' is 100 times smaller than R, the discharge will be 100 times quicker. The number of clock cycles counted during this time interval is given by

$$V - V_{\text{threshold}} = \frac{E_{\text{ref}}}{RC} 100 N_2 T,$$

where T is the clock period.

Second phase

Once, the integrator output equals to the threshold voltage, resistance R is reactivated. A counter starts counting clock pulses from zero. The number of pulses N_3 which have been counted when the integrator output voltage reaches zero is given by

$$V_{\text{threshold}} = \frac{E_{\text{ref}}}{CR} N_3 T.$$

Hence we have the relationship

$$V = \frac{E_X}{RC} N_1 T = \frac{E_{\text{ref}}}{RC} 100 N_2 T + \frac{E_{\text{ref}}}{RC} N_3 T.$$

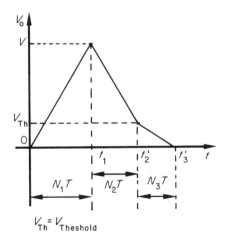

Figure 7.42 Principle of operation of a three fold slope converter

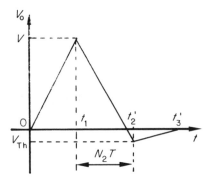

Figure 7.43 Second version of three-fold converter

Therefore

$$E_X = \frac{E_{ref}}{N_1}(100 N_2 + N_3). \qquad (7.6)$$

During this phase, the discharge slows down to allow accurate counting of the last pulses. In comparison with the dual-ramp method, this method produces the same resolution with a shorter conversion time.

A version of this type of converter is to fix the threshold voltage equal to the integrator output at the clock pulse immediately after this output passes through zero (Figure 7.43). At that instant, the number $100 N_2$, which is stored in the pulse counter, gives an over-estimated unknown voltage value E_X. Therefore a reference of $-E_{ref}$ is applied to the integrator input and the resistance R is reactivated. The slope of the output ramp changes polarity and the counter starts counting down from $100 N_2$ until the ramp reaches zero.

7.2.3.5 Quad-slope converter

The basic operation of a quad-slope converter is explained by Figure 7.44.

The integrator has four modes of connection:

— clamped when no conversion is in process;
— grounded input;
— reference input (E_{ref});
— analogue input (E_X).

Voltage $E_{ref}/2$ is continuously applied to the '+' input of the integrator.

After the start pulse is applied, E_{ref} is connected to the input and the integrator output is ramped to the comparator zero crossing. This reset phase has a duration equal to the integrator time constant (RC). Different

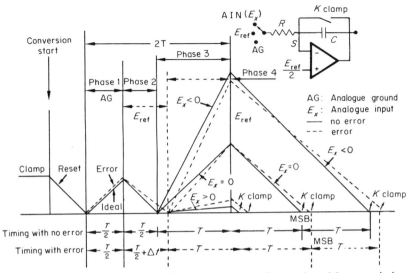

Figure 7.44 Illustration of quad-slope principle. (Reproduced by permission of D. H. Sheingold, Analog Devices, Inc.)

inputs will be applied in sequence to the integrator in order to create four following operations phases.

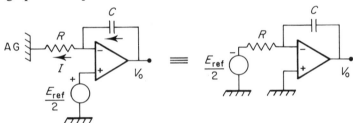

Figure 7.45 Analogue ground is connected to the integrator input

Phase 1. Analogue ground is connected to the integrator input (Figure 7.45). Voltage $-E_{ref}/2$ is integrated for a fixed interval $T/2$. Upon zero crossing of reset phase, phase 1 is initiated and a counter starts counting clock pulses. Recall that $E_{ref}/2$ is applied to the '+' input of the integrator.

Phase 2. E_{ref} is applied to the integrator input. Voltage $(E_{ref} - E_{ref}/2)$ is

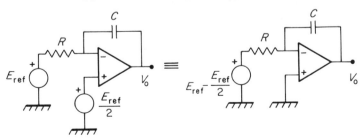

Figure 7.46 E_{REF} is applied to the integration input

integrated; the integrator outputs ramps down until zero crossing is achieved. If there is no offset error, the time for phase 2 will be the same as for phase 1 ($T/2$). Any error will increase or decrease this time by an amount Δt.

Phase 3. E_X is applied to the integrator input. The analogue input (AIN $- E_{ref}/2$) is integrated. The integrator output ramps upwards with a proportional slope. Phase 4 is initiated when the counter that started at the end of the reset phase reaches a count equivalent to $2T$.

At the beginning of phase 3, a second counter is starting at $T \pm \Delta t$.

Phase 4. E_{ref} is applied to the integrator input. Voltage ($E_{ref} - E_{ref}/2$) is again integrated. The integrator output ramps down at the rate $E_{ref}/2$ plus any error until zero crossing once again is achieved. The second counter is stopped. Conversion is not complete. The counter output is a 2s complement representation of the analogue input.

In Figure 7.44, the effect of an error is to shift all crossings by an amount of time ΔT.

If the counter's capacity is $2T$, and if it counts down from all zeroes at the beginning of phase 3, we have at the output of a 13-bit converter the values shown in Table 7.2.

Table 7.2

E_X	bit 12 (sign)			bit 0
$+V_{FS}(1-2^{-12})$	0	1111	1111	1111
0	0	0000	0000	0000
$-V_{FS}$	1	0000	0000	0000

The range indicated in Table 7.2 is recognized as belonging to a 2s complement code. We have supposed that the full-scale voltage is equal to $E_{ref}/2$. In actual practice, in order to avoid negative integration and allow sufficient time after phase 4 begins for offset correction and over-range indication, a somewhat different counting scheme is used, in association with an input full-scale range of $E_{ref}/2.125$, instead of $E_{ref}/2$ as in the Analog Devices 13-bit AD7550.

The counters are then designed in order to give a number corresponding to the converted analogue signal by the relationship

$$N = \frac{E_X}{E_{ref}} 8704 + 4096$$

or

$$N = \frac{E_X}{V_{FS}} 4096 + 4096.$$

Hence the values shown in Table 7.3.

Table 7.3

E_X	N	bit-12			bit-0
$+V_{FS}(1 - 2^{-12})$	8191	0	1111	1111	1111
$+V_{FS} \times 2^{-12}$	4097	0	0000	0000	0001
0	4096	0	0000	0000	0000
$-V_{FS} \times 2^{-12}$	4095	1	1111	1111	1111
$-V_{FS}$	0	1	0000	0000	0000

7.2.4 Principle Specifications of an ADC

Some specifications are identical to those defined in the digital-to-analogue section: for example, resolution, accuracy, offset and gain error, linearity, conversion time, influential factors.

7.2.4.1 Hysteresis error

This is due to a positive feedback of the comparator and must not exceed $\pm q/2$, i.e. half of the input voltage step required to change the least significant bit (q = quantum).

7.2.4.2 Quantization error

The transfer characteristic of an ADC is a stair-case function, as shown in Figure 7.47(a). The quantizating operation introduces an inherent systematic error which can be chosen to be either centred (full line) or default (dotted line).

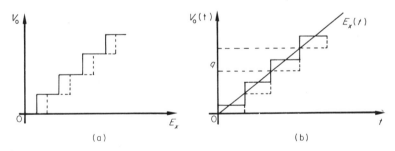

Figure 7.47 ⓐ Transfer characteristic of an ADC; ⓑ quantizating operation of analog signal

For a positive-ramp input voltage E_X as shown in Figure 7.47(b), the quantization error $\varepsilon = V_0(t) - E_X(t)$ can be represented by Figure 7.48. It can be considered like a dynamic noise which adds to the signal and its

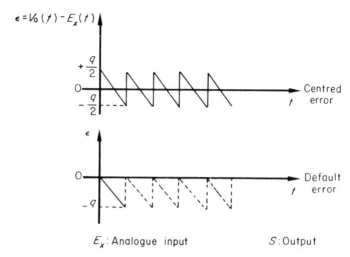

Figure 7.48 Quantization error

root-mean square is:

$$\frac{q^2}{12} \quad \text{for centred error}$$

and

$$\frac{q^3}{3} \quad \text{for default error.}$$

7.2.4.3 Missed codes

It is known that when a differential linearity error is higher than 1 LSB a nonmonotonic response is produced in a DAC. This error causes a missed code which is explained by the example of the following diagram Figure 7.49. When the analogue input is slightly less than the value of 010, it is

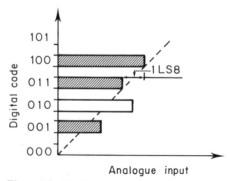

Figure 7.49 Example of missed code: The code 010 doesn't exist

converted to 001, whereas if it is slightly higher than the value of 010 it is converted to 011. Thus the code 010 does not exist.

7.2.4.4 Noise rejection

Noise rejection is an important specification to consider when choosing converters for applications in which the measuring signal is greatly influenced by noise. Two classes of noise may be differentiated.

— series-mode noise, which is superimposed in series with the signal;
— common-mode noise, which exists when floating inputs are used.

The common mode noise is discussed in Chapters 3 and 4.

Series-Mode Noise Rejection

Figure 7.50 shows an example of series-mode noise added to the useful signal. In series with the e.m.f. e generated by the thermocouple hot junction, there is an e.m.f. e_t generated by the relay thermoelectricity contact and also a 50 Hz a.c. voltage e_m induced by the relay coil. Filters can be used to reduce these noise signals. Efficient noise reduction means, the longer the measurements take to perform.

The series-mode rejection ratio is defined as the ratio of, expressed in decibels, the noise peak value to the analogue input signal value that produces the same output change.

Integrating type converters, such as frequency-to-voltage, dual slope, threefold slope, quad-slope, etc., provide rejection of high frequency noise by averaging changes that occur during the sampling period. A fixed averaging period T makes it possible to obtain 'infinite' rejection at frequencies that are integral multiples of $1/T$.

Take an example of measuring a d.c. e.m.f. signal e, which is accompanied by an a.c. noise $V_S \sin(\omega t + \varphi)$. The total signal captured at the input converter is then

$$v(t) = e + V_S \sin(\omega t + \varphi).$$

At the end of the integration interval T, the output voltage of the

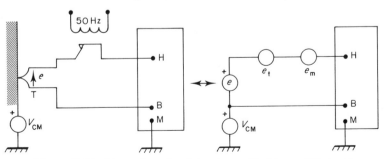

Figure 7.50 Examples of serial noises added to useful signal

integrator will be:

$$\bar{V} = \frac{1}{T}\int_0^T [e + V_S \sin(\omega t + \varphi)]\, dt$$

$$= e - \frac{V_S}{\omega T}[\cos(\omega T + \varphi) - \cos\varphi].$$

In the case that T is a multiple of $2\pi/\omega$, the 2nd term due to the mean noise value over T is equal to zero. Therefore

$$\bar{V} = e$$

The integrator acts like a selective filter at frequencies $\omega/2\pi$ and its harmonics. For other frequencies, its noise reduction is defined by the series-mode noise rejection ratio which is defined as

$$R_{SNR} = \frac{\text{noise peak value}}{\text{error voltage due to interference}}.$$

In this example we have:

$$R_{SNR} = 20\log\frac{\omega T}{\cos(\omega T + \varphi) - \cos\varphi}. \tag{7.7}$$

Figure 7.51 shows the variation of the series-mode noise rejection ratio with noise frequencies. In practical cases, the parasitic signal is essentially a 50 Hz line-frequency pick-up. In order to minimize the effect of this noise, an integrating converter can then be used but the minimum integration time, thus the conversion time, is not less than 20 ms.

For noise frequencies different from $1/T$ and its multiples, the resulting error varies with the phase difference between the signal and the time integration.

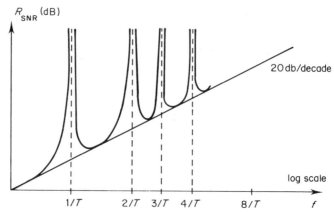

Figure 7.51 Variation of R_{SNR} with noise frequencies

Numerical example Suppose that $T = 20$ ms. The resulting error is indicated in Figure 7.52.
— For a parasitic signal of 25 Hz, maximum error is $2/\pi\, V_s$.
— For a parasitic signal of 75 Hz, maximum error is $2/3\pi\, V_s$.

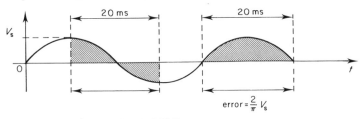

(a) Noise signal of 25 Hz

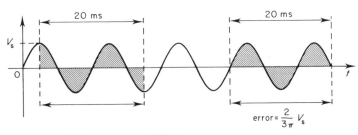

(b) Noise signal of 75 Hz

Figure 7.52 Error varies with phase difference between noise signal and time integration (T) for noise frequencies different from $1/T$ and its multiples

7.2.5 Examples of Applications

7.2.5.1 Structure of a digital filter

A digital filter, which consists of adders, multipliers and delay elements, is used to change a binary input sequence $e(k)$ into a binary output sequence $s(k)$ and thereby implement a desired digital transfer function $F(Z)$. If the input signal is continuous, the number sequence $e(k)$ must first be produced from it. The block diagram of the circuit is shown in Figure 7.53. To fulfill the condition imposed by the sampling theorem, an analogue low-pass filter

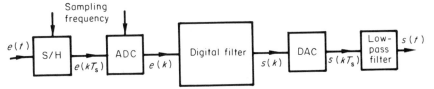

Figure 7.53 Organization of a digital filter

must be used for band limiting. The sample-and-hold circuit that follows takes samples from the band-limited signal at intervals $T = 1/f$. These samples $e(kT)$ are converted to the number sequence $e(k)$ by an analogue-to-digital converter and applied to the input of the digital filter. The output sequence $s(k)$ may be processed digitally, or converted to a continuous output signal by means of a digital-to-analogue converter and a low-pass filter.

The output sequence $s(k)$ is generally computed from the input sequence $e(k)$ by a recurrence expression:

$$s(i) = \sum_{k=0}^{N} e(i-k)h(k), \qquad (7.8)$$

where $h(k)$ is the sampled sequence of the Dirac impulse response of the digital filter.

The factors $h(k)$ required in Eq. (7.8) can be determined from the digital transfer function $F(Z)$, thus allowing the programming of a single arithmetic unit to carry out all the operations needed.

Example The Z-transform of the function $f^*(t)$ shown in Figure 7.54 is given by the following relationship:

$$F(Z) = \mathscr{L}[f^*(t)] \quad \text{with} \quad Z = e^{Tp}.$$

Suppose that $F(Z)$ is given by:

$$F(Z) = \frac{S(Z)}{E(Z)} = \frac{Z^2 + 1}{Z^2 - 2Z + 1}, \qquad (7.9)$$

where $S(Z)$ and $E(Z)$ are respectively Z-transforms of the output and input signals. Then Eq. (7.9) can be written as:

$$Z^2 S(Z) - 2ZS(Z) + S(Z) = Z^2 E(Z) + E(Z). \qquad (7.10)$$

Elsewhere, we have

$$E(Z) = \sum_{k=0}^{\infty} e(k) Z^{-k},$$

$$S(Z) = \sum_{k=0}^{\infty} s(k) Z^{-k}.$$

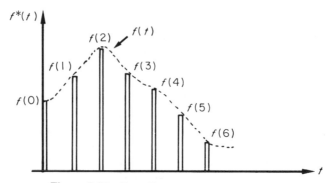

Figure 7.54 Sampling of analog signal $f(t)$

By putting these expressions into Eq. (7.10) and by equalizing the factor of Z^{-k} in the left-hand term to that in the right-hand one, we obtain

$$S(i-2) - 2S(i-1) + S(i) = e(i-2) + e(i).$$

Therefore, we have the recurrence relationship:

$$S(i) = e(i) + e(i-2) + 2S(i-1) - S(i-2).$$

7.2.5.2 Structure of a digital storage oscilloscope

The storage oscilloscope is far from being the answer to all the requirements of everyday use. The digital storage oscilloscope (DSO) lacks the disadvantages of the preceding device, such as its high cost for recording information, its inability to dilate the stored signal for further investigation and to vizualize information preceding the trigger pulses etc. In addition the DSO presents new facilities. For example, digital outputs are available for remote programming and bi-directional data transfer. It also includes standard interfaces for computers, among them the well-known and widely used GPIB/IEE 488 (see Chapter 9) which allows the DSO to be incorporated in an automatic instrumentation system. The functional diagram of a DSO is shown in Figure 7.55.

The input waveform is sampled by means of a sampling unit followed by conversion of the sampled sequence to a number sequence by an ADC. These binary data are then stored in memory under the control of a logic circuit until the memory is full. Once the storing operation is completed, the stored data can then be read out and fed to a DAC followed by a low-pass filter to reconstitute the analogue input signal. The ADC must be a very fast type in order to avoid limiting the upper functional frequency of the DSO.

The horizontal resolution of the DSO depends on the memory capacity. A 4K memory can produce up to 4000 points on the screen, and the horizontal resolution is then 4000 points. If the memory is divided for two separate recordings, the resolution is then also divided by two. A horizontal resolution of N points can be imagined as an elastic string with N pearls placed equidistant from one another. The adjustment of the time base

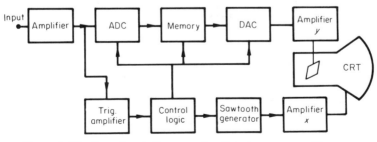

Figure 7.55 Functional diagram of a digital storage oscilloscope

corresponds to the stretching of the elastic string and the distance between pearls. If the elastic string length equals t_{ex} seconds, the time interval represented by the distance between pearls is t_{ex}/N, and all signals which happen during the time interval between two samples are ignored.

For a given horizontal sweep rate, both the bandwidth and the settling time of the DSO are defined by its memory capacity.

(a) The bandwidth is given by the relationship:

$$\Delta f = \frac{F_s}{n},$$

where F_s is the sampling rate and n the number of samples required for a possible reconstitution of the initial sine-wave input signal. (According to Shannon's theorem, a reliable reconstitution requires that F_s must be at least twice the frequency of the input signal.)

While the sweep rate (V) differs from one application to another, the number of points (N) stored in the memory circuit is fixed. The DSO sampling rate is determined by the ratio

$$F_s = \frac{N}{V}.$$

Therefore, for a given sweep rate, the sampling rate, and thus the bandwidth, is proportional to the memory capacity.

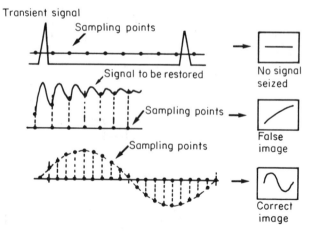

Figure 7.56 Sampling and restoring useful signal

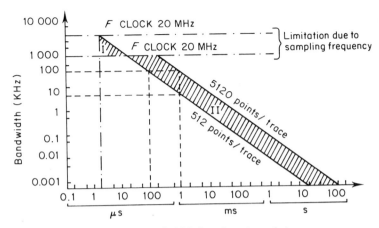

Figure 7.57 Bandwidth is a function of slew rate

It is interesting to note that according to Shannon's theorem the upper limit of the sampling rate determines the DSO bandwidth, but this upper value is reached only for very high sweep rate. Figure 7.57 shows the bandwidth variation as a function of the sweep rate. One can see that for ordinary sweep rates, the DSO bandwidth depends entirely on the memory capacity.

(b) The more important the number of points N is, the faster is the settling time. This proposition can be illustrated by an example shown in Figure 7.58, where the sampled points are situated in the most disadvantageous case with regard to the original signal.

Settling time $T_r = 1.6 \times$ sampling period:

$$T_r = 1.6 \frac{1}{F_s} = 1.6 \frac{V}{N}.$$

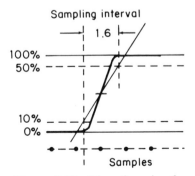

Figure 7.58 Rise time in the most unfavourable case

The major advantage of the DSO is that when the stored signal includes rapid phenomena, a horizontal expansion permits examination of these phenomena in detail. This is because a large memory capacity and a associated increase in expansion is equivalent to the magnifying effect of a dual time-base oscilloscope. Besides, the DSO can be used in a real-time system, where it can give great help to the process controls.

Chapter 8
SAMPLE-AND-HOLD CIRCUITS-MULTIPLEXERS

8.1 SAMPLE-AND-HOLD CIRCUITS

8.1.1 Principle of Operation

A sample-and-hold (S/H) circuit has an analogue input terminal, an analogue output terminal, and a digital control input. Its basic elements are a storage capacitor, a switch and its drive-circuitry. A simplified schematic diagram is shown in Figure 8.1. There are two modes of operation:

— During sample or track mode, the output follows the input voltage as faithfully as possible and usually with unity gain.
— During hold, the output ideally stores the last value it had when the command to hold was given, and this value is retained until the logic input dictates sample. At this time the output ideally jumps to the input value and follows the input until the next hold command is given.

These two modes are programmed by digital levels of the control input, commonly logic 1 for sample and logic 0 for hold.

8.1.2 Choice of sampling frequency

In ADC systems the sample-and-hold circuit finds two common applications. First, to sample during a short time interval the value of a varying analogue voltage and then hold it constant for the duration of the conversion. Second, in conjunction with analogue multiplexers, to sample the multiplexed analogue voltage while the multiplexer is acquiring the next analogue channel. In practice, the input signal is sampled at equidistant instants $t_i = iT_s$, where $F_s = 1/T_s$ is the sampling rate. Obviously the sampling rate is one factor along with the ADC conversion time, the input voltage rate of change, etc., which determines the accuracy of the recovered signal.

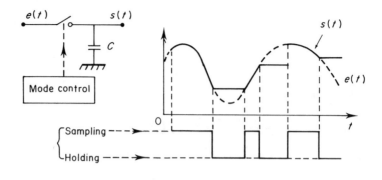

$e(t)$: input signal $\qquad$ $s(t)$: output signal

Figure 8.1 Principle of operation of a sample–hold device

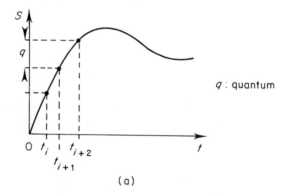

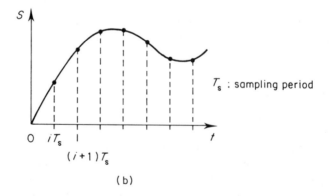

Figure 8.2 Quantizing and sampling operation: (a) quantizing of signal $s(t)$; (b) sampling of signal $s(t)$ with sampling period T_s

8.1.2.1 Sampling theorem

In order to obtain a simple mathematical description, the sampled values in Figure 8.2 are considered as a result of a product of the input signal with a series of Dirac impulse functions, as illustrated in Figure 8.3:

$$s^*(t) = s(t) P_{T_s}(t), \qquad (8.1)$$

where the series of Dirac impulse functions can be written as:

$$P_{T_s}(t) = \sum_{k=-\infty}^{+\infty} \delta(t - kT_s).$$

For an investigation of the information contained in the impulse function sequence represented in Eq. (8.1), we consider its spectrum. By applying the Fourier transformation to Eq. (8.1), we obtain

$$s^*(f) = \frac{1}{T_s} \sum_{k=-\infty}^{+\infty} s\left(f - \frac{k}{T_s}\right)$$

where $s(f)$ is the Fourier transformation of the input signal $s(t)$.

It can be seen that this spectrum is a periodic function, the period being identical to the sampling rate F_s. When Fourier-analysing this periodic function, it can be shown that the spectrum $s^*(f)$ is, for $-F_c \leq f \leq F_c$, identical to the spectrum $s(f)$ of the original wave-form. Thus it still contains all the information although only a few values of the function were sampled. The only restriction is explained with the help of Figure 8.4. The original spectrum reappears unchanged only if the sampling rate is chosen such that adjacent bands do not overlap. According to Figure 8.4, this is the case for $F_s \geq 2F_c$, this condition being known as the sampling theorem. An example of spectrum folding is shown in Figure 8.5.

This involves the precondition that the input function $s(t)$ is band-limited, i.e. that its spectrum $s(f)$ above a frequency F_c is, at least approximately, zero. If this condition is not fulfilled, it is often possible to enforce it by using a low-pass filter at the input without unduly changing the signal.

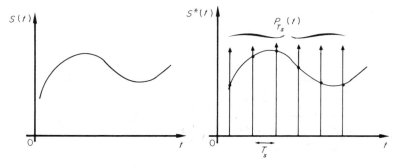

Figure 8.3   Model of sampled signal $s(t)$

250

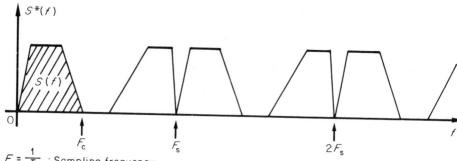

$F_s = \frac{1}{T_s}$: Sampling frequency
F_c : Maximum frequency of $S(f)$ spectrum

Figure 8.4 Spectrum of $S^*(t)$ in the case of $F_s \geq 2F_c$

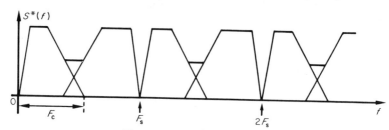

Figure 8.5 Spectrum folding

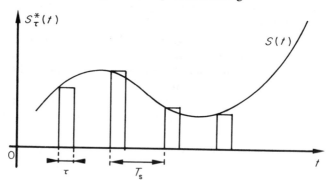

T_s : Sampling period τ : Holding time interval

Figure 8.6 Sampled and held signal

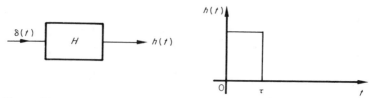

Figure 8.7 Dirac impulse is approximated by a finite amplitude and a finite time interval

8.1.2.2 Sample-and-Hold (Figure 8.6)

As a real system is unable to generate Dirac impulse functions, the impulses must thus be approximated as in Figure 8.7 by a finite amplitude and a finite time interval. The sampled values are thus held constant during a time interval τ as illustrated in Figure 8.6.

The effect is as if the sampled signal $s^*(t)$ pass through an additional system which produces a rectangular $h(t)$ output in response to a Dirac impulse $\delta(t)$ input.

The Fourier transformation yields the spectrum:

$$S_\tau^*(f) = H(f)s^*(f).$$

with

$$|H(f)| = \left|\tau \frac{\sin \pi f \tau}{\tau f}\right|,$$

the Fourier transformation of the additional system, which is the same as for the Dirac impulse sequence, except for a superimposed weighting function causing an attenuation of the higher-frequency component.

The staircase function of Figure 8.6 is particularly interesting when the pulse width τ is identical to the sampling period T_s, as shown in Figure 8.9. The spectrum is then given by

$$s_{T_s}^*(f) = T_s \frac{\sin (\pi f T_s)}{\pi f T_s} s^*(f).$$

The magnitude of the weighting function is represented in Figure 8.10, along with the symbolic spectrum of the resulting signal $s_{T_s}^*(t)$. At half of the sampling rate, an attenuation of 0.64 is obtained whilst at $0.2F_c$ it is only 0.94. The influence of the weighting function on the spectrum below the band limit F_{max} thus remains negligibly small as long as $F_c \simeq 5F_{max}$.

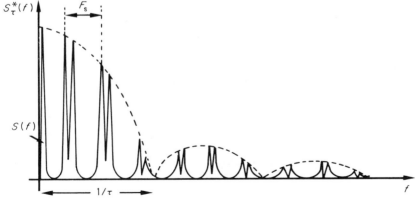

Figure 8.8 Spectrum of a sampled and held signal

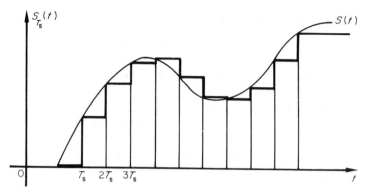

Figure 8.9 An interesting case is produced when holding time interval is equal to sampling period

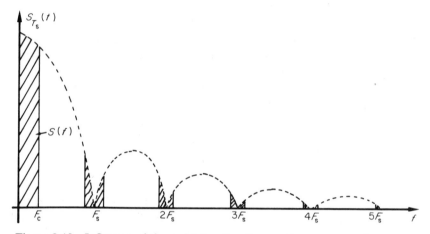

Figure 8.10 Influence of the weighting function on the spectrum when $\tau = \tau_s$

The recovery of the original signal requires a low-pass filter which stops all frequencies in the spectrum above F_{max}. It is even advantageous to use a staircase function instead of the Dirac impulse functions, as the resulting weighting function itself has low-pass filter characteristics.

The deformation of the amplitude spectrum in the pass-band region can be compensated by increasing the gain of the low-pass filter near the band limit. For a sufficient decrease in attenuation above F_c, the insertion of a zero in the frequency response, at $F_s - F_c$, is recommended.

8.1.2.3 Relationship between ADC resolution and sampling rate

Suppose that an input ramp voltage, as shown in Figure 8.11, is to be converted:

$$s(t) = Vt,$$

where V is the slope of the ramp.

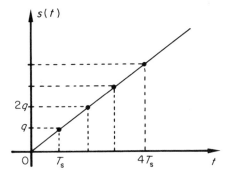

Figure 8.11 Determination of sampling frequency and ADC resolution

The signal change between two consecutive samples is less than the resolution q of the converter if the following condition is satisfied:

$$q \geq VT_s, \quad \text{i.e.} \quad F_s \geq V/q.$$

It follows that, for a signal having a maximum rate of change V_{max}, the sample rate must verify:

$$F_s \geq \frac{V_{max}}{q}$$

Numerical example Determine the minimum sampling rate to convert a sinusoidal input signal $V_{in}(t) = 10 \sin 2\pi 100 t$, with a 10-bit ADC, in order that the change between two consecutive conversions shall not exceed 1 LSB.

The highest rate of change of the sine wave is at phase angles of 0° and 180°, that is:

$$V_{max} = 10 \times 2\pi 100 \text{ V/s}.$$

1 LSB is approximately $\dfrac{10}{1024} \simeq 0.01$ V.

Thus, the sampling rate must verify the condition:

$$F_s \geq \frac{10 \times 2\pi \times 100}{0.01}, \quad \text{i.e.} \quad F_s \geq 6.28 \times 10^5 \text{ Hz}.$$

F_s must be 3000 times as high as the frequency imposed by Shannon's theorem.

This method of sampling rate determination is specially interesting for restoring transient phenomena. In general, it is possible to pick out two principal cases (Figures 8.12 and 8.13).

(1) When the useful signal spectrum is known, the sampling frequency may depend only on the highest frequency F_c (Figure 8.13).
(2) When the signal spectrum is unknown (Figure 8.12) and the signal must be restored point by point at equidistant instants with a required

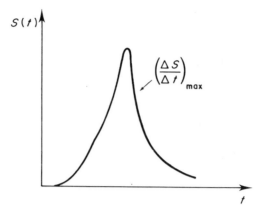

Figure 8.12 Transient phenomena

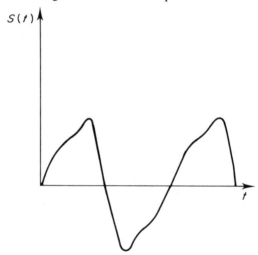

Figure 8.13 Band-limited signal

accuracy, the sampling frequency can be chosen by the consideration of the highest rate of change $V_{max} = (\Delta s/\Delta t)_{max}$.

Example Assume that we have to restore a short-duration transient voltage varying between 0 and 10 V, with a limiting error equal to 40 mV at every sampled point. Determine the ADC resolution and calculate the sampling frequency. The highest rate of change $(\Delta s/\Delta t)_{max}$ is equal to 4000 V/s.

Solution

ADC resolution is $10\,\text{V}/2^n = 40\,\text{mV}$ so that 8-bit ADC can be used.

$$\text{Sampling frequency} = F_s \geq \frac{4000}{40 \times 10^{-3}} = 100\,\text{kHz}.$$

If the sampling frequency is smaller than 100 kHz, an error greater than the allowed limiting value can be produced in places.

8.1.3 Principal Characteristics of a Sample-and-hold Amplifier

In practical realizations, sample-and-hold circuits do not meet the two ideal modes of operation, which are described in the introduction, due to the influence of external sources and loads. The combination of the sample-and-hold capacitor and the source resistance acts like a low-pass filter. In order to provide rapid charging of this capacitor, the source resistance must be kept low.

Usually, a data-acquisition sample-and-hold circuit must exhibit ultra-fast signal acquisition, with minimal droop for a matter of 1–50 μs. As the sample-and-hold capacitor is charged through the input source resistance with time constant CR, the smaller the value of the resistance R, the quicker is the charging.

By contrast, an output data distribution sample-and-hold circuit can be updated in a more leisurely manner, but it is often required to 'hold' an analogue value for many milliseconds—or even seconds—in the interval between updates. To minimize the discharge current, a high-quality capacitor with low dielectric absorption and a large value load resistance are required.

For these reasons, a sample-and-hold circuit usually includes input and output buffer-amplifiers. However, during sample-to-hold, hold, and hold-to-sample states, the dynamic nature of the mode-switching introduces a number of specifications peculiar to sample-and-hold amplifiers. The most important of these are defined below. They include the aperture time and its uncertainty, charge transfer, feedthrough and droop and acquisition time.

8.1.3.1 Sample state

During the sample mode, the output must follow the input as faithfully as possible. Figure 8.14 represents the ideal transfer characteristic and the real transfer characteristic respectively by a dotted line and a full line.

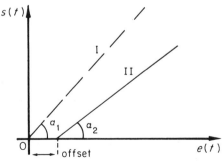

I : Ideal transfer characteristic
II : Actual transfer characteristic

Figure 8.14 Zero offset and gain offset

Zero offset and gain offset result from imperfections of switch and amplifier. Some products are equipped with a compensation circuit to minimize these errors.

8.1.3.2 Hold state

During the hold mode, errors result essentially from the two phenomena of feedthrough and droop illustrated in Figure 8.15.

Feedthrough is the fraction of the input signal variation or a.c. input wave-form that appears at the output when the sample-and-hold amplifier is in the hold mode. It is caused by stray capacitive coupling from the input to the storage capacitor, principally across the open switch. The feedthrough rejection value varies from about -90 dB for low frequency (Hz) to -60 dB for high frequency (1 MHz).

The droop is the change of the output voltage during hold mode as a result of FET switch leakage currents and output amplifier bias current through the storage capacitor. Its polarity depends on the sources of leakage current within a given device. Since drive current is finite, and leakage current in hold is not zero, the capacitance (if large) limits the slewing rate in the sample state and (if small) converts leakage currents to 'droop' in the hold state. In sample-and-hold modules, the capacitance is usually fixed, and the properties of the complete device are optimized for one condition and so specified. In sample-and-hold monolithic ICs, the capacitor is omitted, and furnished by the user both for flexibility and because good capacitors for this purpose are difficult to integrate. The optimum capacitance can be selected for the specific application by using design curves of-performance of capacitance given on some IC data sheets. In ICs, the droop is specified as a droop current whereas in modules it is a droop rate dV/dt.

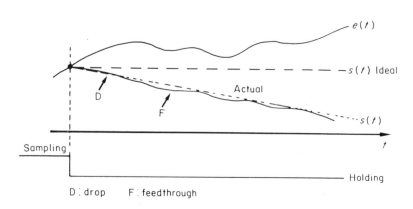

Figure 8.15 Feedthrough and droop

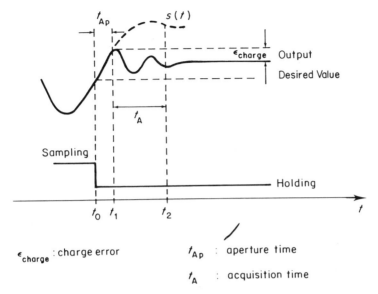

Figure 8.16 Sample-to-hold state

8.1.3.3 Sample-to-hold state (Figure 8.16)

An important characteristic in this mode is the aperture time t_{Ap}. It defines the delay t_1 between switching off the control voltage to, and the actual cut off of the series switch. As the sample is, in effect, delayed by this interval, the hold command would have to be advanced by this amount for precise timing. The delay $(t_1 - t_0 = t_{Ap})$ is subject to a certain variation known as aperture jitter, Δt_{Ap} which defines the degree of uncertainty as to the beginning of the hold mode.

Numerical example Suppose that a signal, whose maximum rate change is equal to 1 V/μs has to be sampled using a sample-and-hold with an aperture time jitter of ±10 ns. The maximum error resulting from this uncertainty is given by

$$\varepsilon = \frac{1}{10^{-6}} \times 10 \times 10^{-9} = 10^{-2} \text{ V}.$$

If the full-scale voltage is 10 V, the precision is then limited to 0.1%.

After the switch is fully opened, a time interval $(t_2 - t_1)$ is required for the output to attain its final value within a specified fraction of full scale. This time interval is called the settling time.

An additional source of error is that, on switching off, some charge is taken from the storage capacitor C by the finite gate-to-drain capacitance of the FET switch. This produces a voltage error called sample-to-hold offset or charge offset.

$$\Delta V_0 = \frac{C_{GD}}{C} \Delta V_c$$

As the gate-to-drain capacitance C_{GD} is in the order of several picofarads, a storage capacitance of at least 1 nF is required for an accuracy of 0.1%.

Numerical example The AD 583 from Analog Devices is a monolithic sample-and-hold circuit with a maximum charge transfer of 20 pC. If an external holding capacitor of 300 nF is used, then the sample-to-hold offset will be:

$$\varepsilon = \frac{20 \times 10^{-12}}{300 \times 10^{-9}} \simeq 7 \times 10^{-5} \, \text{V}.$$

8.1.3.4 Hold-to-sample state

An important factor determining the quality in this mode is the acquisition time t_{AC}. It defines the time required in the worst case to charge the capacitor to a given percentage of the input voltage. This includes switch delay time, slewing time and settling time for a given output voltage change.

Example The acquisition time of the AD 582 is:
— 6 μs for a 10 V step to 0.1% with $C = 100$ pF
— 25 μs for a 10 V step to 0.01% with $C = 1000$ pF.

Figure 8.17 illustrates various dynamic characteristic of the sample-and-hold circuit.

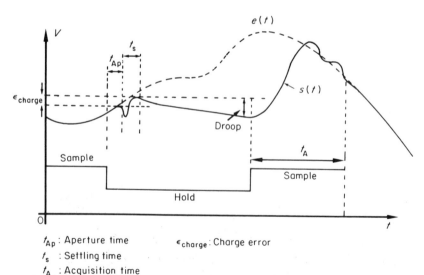

t_{Ap} : Aperture time ε_{charge} : Charge error
t_s : Settling time
t_A : Acquisition time

Figure 8.17 Various dynamic characteristics of a sample–hold device

8.1.4 Practical Realizations

Commercial sample-and-hold circuits differ in their internal structure. Some are designed to be fast, some to be accurate and some to realize a suitable compromise between these two characteristics.

8.1.4.1 Cascade configuration

Figure 8.18 shows the simplest sample-and-hold circuit where the input and output signals are isolated from the switch and the storage capacitance by two voltage followers.

The first amplifier serves to reduce the acquisition time if the signal source has a high impedance. Depending on the required acquisition accuracy,

$$t_{Ac} = R_{ON}C \times \begin{cases} 4.6 & \text{for} \quad 1\% \\ 6.9 & \text{for} \quad 0.1\%. \end{cases}$$

The device is fast rather than accurate because of the amplifier's imperfections. In particular the error due to the summing of the offset voltages of the two cascading amplifiers may be troublesome.

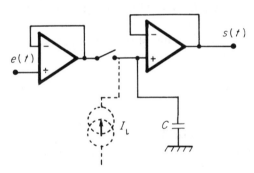

Figure 8.18 Sample–hold circuit: cascade configuration

Example of IC realization
LH 0043 from National Semiconductor
— Offset voltage adjustable by a 10 kΩ potentiometer.
— Aperture time about 20 ns.
— Acquisition time: 4µs for a 5 V step.
— Leakage current: $I_{leakage} = 10$ pA at $T = 25°C$.

Note that the storage capacitance value is determined by the prescribed maximum droop rate

$$C = \frac{i_{leakage}}{\text{Droop rate}}.$$

If a 5 mV/s droop rate is desired, the required capacitance will be

$$C = \frac{10 \times 10^{-12}}{5 \times 10^{-3}} = 2000 \text{ pF}.$$

8.1.4.2 Feedback configuration

Offset voltage and common-mode effects can be eliminated by employing feedback across the entire arrangement, including the first amplifier, as in Figure 8.19. If the switch is on, the output potential of the first amplifier assumes a value such that

$$V_0 = A(V_0 - e)$$

Therefore

$$V_0 = e \frac{A}{A-1}.$$

With a high value of gain A, we can assume that the output follows the input, $V_0 = e$. Thus, the offset errors which may arise from the two amplifiers or from the switch are eliminated. However, the acquisition time is sacrificed for the benefit of the gain in accuracy.

During the hold state, the input amplifier is necessarily saturated, due to the difference between the voltage applied to the inverting input, which remains constant at the output of the sample-and-hold circuit, and the changing signal applied to the non-inverting input. The time for recovery after saturation is large and the acquisition time may attain many tens of microseconds. One solution to prevent the first amplifier from becoming saturated is shown in Figure 8.20, where two more switches have been added. In the hold mode, the input amplifier is disconnected from the circuit by closing K_2 and opening K_3, thus eliminating its saturation.

Figure 8.21 shows another solution by using diodes. During the track mode, these two diodes are non-conducting and the whole circuit behaves as an amplifier follower. When the switch is opened, one of these two diodes is forward-biased and the first amplifier loop is closed.

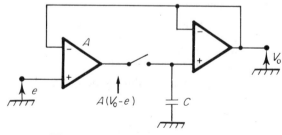

Figure 8.19 Feedback configuration

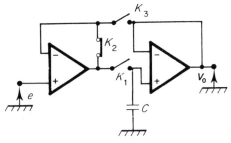

Figure 8.20 Elimination of saturation of the first amplifier

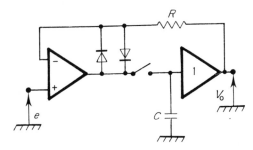

Figure 8.21 Elimination of saturation of the first amplifier using diodes

Examples of industrial realizations
— From National Semiconductor:
  LF 198 – LF 298 – LF 398
— From Datel:
  SHM LM2.

8.1.4.3 Sample-and-hold circuit with integrator

Instead of a grounded capacitor with a voltage follower, an integrator can be used as the analogue storing element.

Figure 8.22 shows this version. In order to prevent the first amplifier

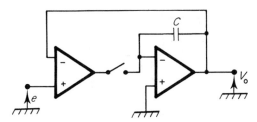

Figure 8.22 Sample–hold circuit with integration

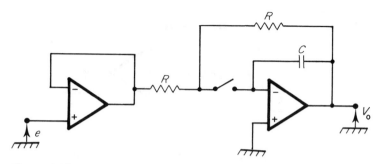

Figure 8.23 Integration configuration without saturation of the first amplifier

becoming saturated in the hold mode the series switch is connected to the summing point and is therefore easy to control. The circuit shown in Figure 8.23 can be adapted.

This family of sample-and-hold amplifiers is among the fastest modules.

Example of industrial realization
AD SHM 5 from Analog Devices.
— Acquisition time: 250 ns to 0.01%.

8.1.5 Application of a Sample-and-hold in a Transient Signal-measuring System

As RAM technology presents a relatively low access time and writing cycle, the association of a RAM memory with a sample-and hold amplifier followed by an ADC makes the acquisition of rapid transient signals possible. Figure 8.24 shows a transient recorder. The one-shot occurrence analogue signal is first converted to binary and stored in a RAM memory. Upon demand, these data can be read out and converted to an analogue signal.

The measuring system begins to record as soon as the transient signal $s(t)$

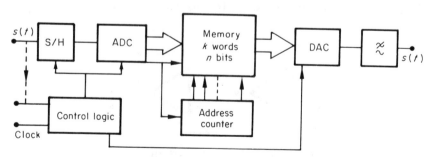

Figure 8.24 Transient signal-measuring system

appears at the sample-and-hold input. The signal is sampled and converted to digital data, which is stored in RAM memory. The memory address register is incremented by a counter after each sample has been stored. After the memory is full, the signal can be read out and converted to an analogue signal by a DAC followed by a low-pass filter. As the signal is recorded during KT_s where K is the memory capacity and T_s the sampling period, one can choose T_s according to the signal duration and a desired accuracy.

8.2 ANALOGUE MULTIPLEXERS

8.2.1 Definition

An analogue multiplexer is composed of several switches that in their on condition transmit analogue signals without attenuation or distortion. Conversely, in their off condition they represent, ideally, an infinite impedance. These switches are actuated (closed) one at a time to connect successively many analogue signal sources to a single output. Analogue multiplexers used to have 4–16 single-ended inputs and one output. A block diagram of an analogue multiplexer is shown in Figure 8.25. These switches are driven by the binary decoder outputs and an 'enable' signal.

Some modules have an incorporated buffered amplifier at the output in order to minimize the resistive loading effects of the source resistance and multiplexer on resistance. It is quite obvious that the switch performances greatly affect the overall performances of the device. For instance, in the absence of output buffering, one must be aware that the multiplexer leakage current could be non-negligible, for it is the sum of $(N - 1)$ leakage currents flowing from 'open' switches. The gate-to-source and gate-to-drain capacitance of the CMOS FET switches, combined with the RC time constants of the source and the load determine the settling time of the multiplexer. As governed by the charge transfer relation $i = C\,dv/dt$, the charge currents transferred to both load and source by switches are determined by the amplitude and rise time of the signal driving the CMOS FET switches and

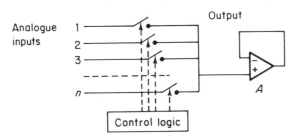

Figure 8.25 Analogue multiplexer

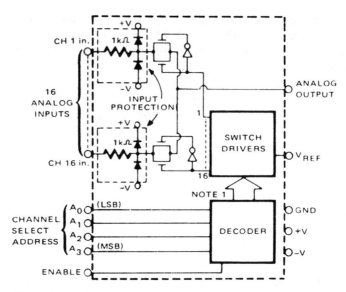

Figure 8.26 Functional block diagram of MPC 16S multiplexer from Burr and Brown

the gate-to-drain and gate-to-source junction capacitances. The junction capacitances of the off channels, which are additive, generate cross talk from the off channels to the multiplexer output.

The selected channel depends on the state of n binary address lines and an 'enable' input. Each switch is driven by one of N decoder outputs. When the address change occurs, two switches may be simultaneously closed during the transition (about $0.3\,\mu s$ for CMOS devices). To avoid that undesirable occurrence, the 'enable' input can be used to validate the output data of the decoder only when it has settled down, i.e. for example $0.5\,\mu s$ after the change occurs. Another solution is to design a circuit such that the switches change state from opened to closed quicker than from closed to opened.

Monolithic CMOS analogue multiplexers with up to 16-channel input, self-contained output buffering and binary channel address decoding are now available (for example, the Burr–Brown CMOS Analog multiplexer MPC 16 S, Figure 8.26).

8.2.2 Channel Expansion

As a great many channels are required for industrial data acquisition systems, several analogue multiplexers must be connected together to increase the channel number. Traditionally, there are two expansion modes. Figure 8.27 shows the first of these. It is called single-mode expansion, for M multiplexers of N channels are connected to a single node to allow

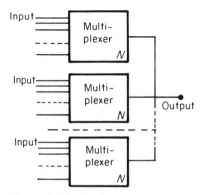

Figure 8.27 Single-mode expansion

multiplexing of $M \times N$ channels. Another possibility is the two-tiered structure shown in Figure 8.28, where M outputs of an N-channel multiplexer are connected to M inputs of an M-channel multiplexer. The total number of channels is then $M \times N$. This second solution has several advantages over the first one. While the programming of the single node expansion requires an M-decoder to enable one out of M parallel multiplexers, the two-tier expansion does not. Moreover, the two-tier configuration offers the added advantages of reduced off channel current leakage (thus reduced *offset*) from $MN - 1$ to $(M - 1) + (N - 1)$, reduction of off channel parasitic capacitance from $(MN-1)C_{DG}$ to $[(M-1)+(N-1)]C_{DG}$, better CMR, and a more reliable configuration if a channel should fail in the on condition (short). Should a channel fail while in the on condition in the single-node configuration, data cannot be taken from any channel, whereas in the multi-tiered configuration only one channel group will fail for the same cause (4 to 16 channels are concerned).

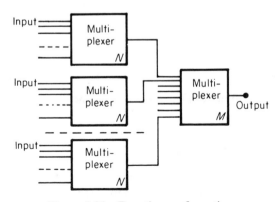

Figure 8.28 Two-tier configuration

Numerical example To realize a 64-channel multiplexer, 9 multiplexers of 8 channels are required for a two-tier structure against 8 multiplexers for one node expansion. But there are only 14 opened switches in the two-tier expansion and 63 in the single node.

The multi-tiered configuration is also used to increase the switching rate. Consider a multiplexer with a settling time of $t\,\mu s$. As the channel N can be activated only when the channel $(N-1)$ is off, the switching time is then $2t\,\mu s$ (Figure 8.29). To reduce this wasted time, either the fast and costly multiplexers or the association of two slow but high-channel number multiplexers with a fast two-channel multiplexer in a two-tier structure can be used.

While the third fast multiplexer connects one channel of the first group, for example, to the output, the transition can take place in the second group or conversely. Therefore, a doubled switching rate can be reached (Figure 8.30).

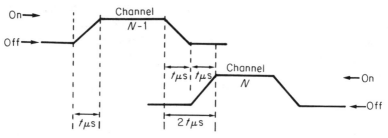

Figure 8.29 Timing diagram of two successive switching operation

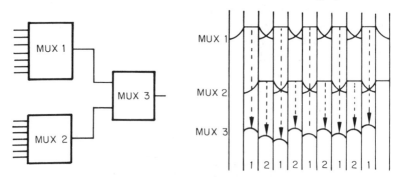

Figure 8.30 Two-tier configuration allowing increase in switching rate

8.2.3 Differential Multiplexing

When multiplexing of low-level data is involved, a combination of a differential multiplexer and an instrumentation amplifier can be used as shown in the block diagram of Figure 8.31. Each differential channel

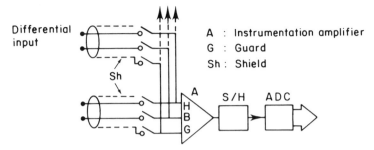

Figure 8.31 Differential multiplexer

multiplexer usually has two inputs. To improve common-mode rejection, some multiplexers present an additional input for each channel. This is reserved for the guard of the individual differential inputs which should be connected to the instrumentation amplifier shield.

8.3 USING A MULTIPLEXER AND SAMPLE-AND-HOLD AMPLIFIER IN A DATA-ACQUISITION SYSTEM

Figure 8.32 shows a data-acquisition system in which all the multiplexer inputs share the same sample-and-hold and ADC.

Different operation phases occur as follows:

Phase 1. Channel Selection The signal from the selected channel appears at the multiplexer output after a certain settling time t_s.

Phase 2. Sampling. The selected signal (channel i) is sampled during the acquisition time interval t_a.

If $t_a \ll t_s$, the sample command is given after the channel selection is completed.

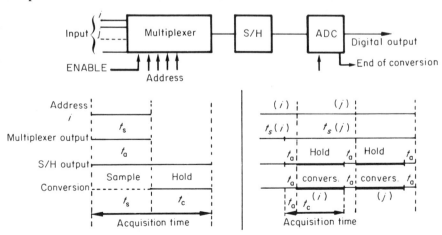

Figure 8.32 Operation of a multiplexer and a sample–hold circuit in a data acquisition system

If $t_a \simeq t_s$, the sample command is given nearly simultaneously with the channel selection, with only a small appropriate delay.

Phase 3. Hold and conversion. The selected analogue signal is now converted into digital data by the ADC. This phase stops when the ADC gives an 'end of conversion' output.

The acquisition time of the system is then $t_s + t_a + t_c$.

For most efficient use of time, overlap programming can be used. Thus the multiplexer seeks the next channel to be converted while the sample-and-hold, in hold mode, is having its output converted. When conversion is complete, the status line from the converter causes the sample-and-hold to return to sample and acquire the next channel. When the acquisition time is over, either immediately or upon command, the sample-and-hold is then switched to hold, a conversion begins, and the multiplexer switch moves on.

Using this approach the overall acquisition time is approximately reduced to the conversion time t_c, because the acquisition time of the sample-and-hold is usually negligible compared with t_c.

8.4 DIGITAL MULTIPLEXER

8.4.1 Principle of Operation

It is often necessary to sample the logic states of different variables one after the other and relay them to a common output. In such cases, a multiplexer as shown in Figure 8.33 is employed. Depending on the state of n address inputs (data select inputs), the output is connected to one of the second data inputs.

The circuit shown in Figure 8.34 is constructed in such a way that when the "enable" input is at logic 1, the output is linked to the input, the index

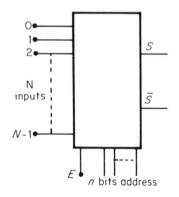

Figure 8.33   Digital multiplexer

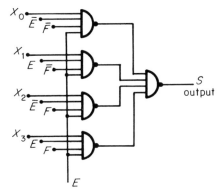

Figure 8.34 Four inputs multiplexer

number of which equals the binary number defined by the two data-select variables E and F. As can be seen from Figure 8.34,

Enable = 1 $S = (X_0 \bar{E}\bar{F} + X_1 E\bar{F} + X_2 \bar{E}F + X_3 EF)$

Enable = 0 $S = 0$

It follows that the Boolean product of the data-select variables is 1 only for the input variable for which the index number corresponds to the chosen value. If, for instance, $F = 1$ and $E = 0$,

$$S = X_0 \times 1 \times 0 + X_1 \times 0 \times 0 + X_2 \times 1 \times 1 + X_3 \times 0 \times 1 = X_2.$$

Figure 8.35 illustrates a block diagram of another kind of multiplexer, the output of which is not a single bit but an x-bit word, at the input there are N words of x bits.

One of the digital multiplexer applications is the parallel-to-serial

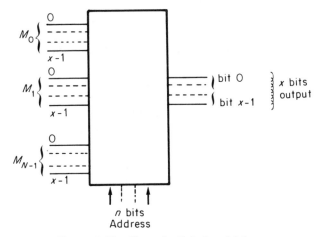

Figure 8.35 N words digital multiplexer

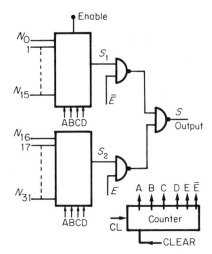

Figure 8.36 Example of parallel-to-series conversion

conversion, whose functional principle is illustrated in Figure 8.36. In this example, after a 32-bit word is fed in parallel to the inputs of two digital multiplexers, each of these inputs is selected by the states of the binary counter outputs: A, B, C, D, E, $\bar{E}$. A clock generator is used to increment the counter. For instance, the 17th bit will appear at the output when the binary counter outputs satisfy $A = 1$, $B = C = D = 0$, $E = 1$.

8.4.2 Using the Digital Multiplexer in a Data-acquisition System

Although the conventional way to digitize data from many analog channels is to introduce a time-sharing process, whereby the input of a single ADC is multiplexed in sequence among the various analogue sources, an alternative parallel conversion approach is becoming increasingly practicable. The cost of ADCs has dropped radically during the past decade, and it is now possible to assemble a multi-channel conversion system, with one converter for every analogue source, just as economically as the conventional analogue multiplexed system (Figure 8.37).

There are a number of important advantages to this parallel conversion approach, which is virtually standard practice for resolver/synchro conversions beyond about the 10-bit accuracy level. First of all, it is obvious that slower converters may be used to obtain a given digital throughput rate. For a constant data rate, however, with more channels (and fewer conversions per channel), the reduced conversion speed, plus the fact that each converter is looking at continuously changing data rather than jumping from one level to another, may allow the sample-and-holds to be eliminated. These factors may result in significant cost saving.

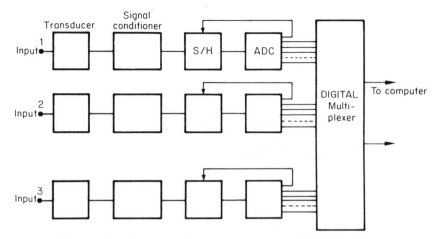

Figure 8.37 Data acquisition system using a digital multiplexer

The parallel conversion approach provides a further advantage when applied to industrial data-acquisition systems, where many strain gauges, thermocouples, thermistors, etc. are strung out over a large geographical area. In essence, by digitizing the analogue signals right at their source and transmitting back to the data centre serial digital data rather than the original analogue signals, a considerable immunity to line frequency (50–60–400 Hz) pick-up and ground-loop interference is achieved. Among other factors, the digital signals can be easily transmitted by opto-electronic means which provide a perfect isolation between transmitter and processing instrument.

Chapter 9
DATA ACQUISITION AND PROCESSING

9.1 DATA-ACQUISITION SYSTEMS

9.1.1 Building up a Data-Acquisition System

Electronic component manufacturers offer a range of modules with a variety of performances and costs. To set up a data-acquisition system, the user should know exactly the characteristics of the desired system, in order to choose the right components in a discerning way. Among the important parameters which should influence the designer, we may select the following essential points:

(1) analogue measurement acquisition;
(2) amplification;
(3) filtering;
(4) sampling rate;
(5) conversion;
(6) transmission.

9.1.1.1 Analogue measurement acquisition

Metrological performance attributes of a system (resolution, accuracy, stability, etc.) are mostly defined by its design. A problem which needs particular care is, without doubt, measurement of low-level signals in a strongly contaminated industrial environment. In fact, the latter is often responsible for large interferences, from many and various origins such as electrostatic (capacitive coupling) and electromagnetic (inductive coupling). On these are superimposed common-mode voltages, as in general signals originating from sensors have to be amplified by a differential device (instrumentation amplifier, isolating amplifier, etc.). To preserve the integrity of the measures, effective shielding is required. The designer must pay attention to wiring by locating the low-level measuring wires remotely from the power cables and by using magnetic and electrostatic shields, twisted pair wires and so on. It can be shown that braided shields reduce electrostatic noise by 49 dB whilst twisted wires with 2.5 cm pitch reduce electromagnetic noise by 43 dB. Using one amplifier on each measuring channel may be quite expensive if there are many inputs. So the amplifier is in general put behind the multiplexer, especially in rapid acquisition systems.

9.1.1.2 Amplification

On account of the diversity of sensors and the fixed input voltage range of ADCs, it is often mandatory to use variable-gain amplifiers. Two methods are used:

— *External programming of gain.* As the computer sets the gain of each measurement channel, the user must specify in advance the amplitude of signals delivered by the sensors, in order to detail the program statements which set these gains.
— *Automatic programming of gain.* Gain control is elaborated within the measuring chain, as a function of input signal amplitude. The latter may be quickly detected by a flash converter, which provides then a digital signal for controlling the amplifier gain. But this method is rarely used in rapid systems.

9.1.1.3 Filtering

If the scanning rate is $f = 1/T$, and the number of distinct channels is n, the input spectra being limited to f_{max}, then Shannon's theorem leads to the condition:

$$nf_{max} < f/2$$

assuming that the n channels are scanned sequentially with period nT.

Numerical example For $f = 100$ kHz and $n = 256$ channels, the input spectra must be limited to

$$f_{max} \leq \frac{100 \text{ kHz}}{2 \times 256} \simeq 200 \text{ Hz}$$

Practical signals show very different spectra. Generally, those having spectra with wider frequency range (for example: non-sinusoidal periodic signals) must be sampled more often than the ones whose spectra are more restricted.

Example Assume signals belonging to three groups n_1, n_2, n_3, whose spectra are limited to f_{1max}, f_{2max} and f_{3max}. ($f_{1max} < f_{2max} < f_{3max}$, $n_1 + n_2 + n_3 = n$).

If scanning is sequential, this is equivalent to scanning a number of channels N:

$$N = n_1 + \frac{f_{2max}}{f_{1max}} n_2 + \frac{f_{3max}}{f_{1max}} n_3$$

whose spectra are all limited to f_{1max}.

In other words, during one period NT, each channel of group n_1 will be scanned once. Each channel of group n_2 will be scanned f_{2max}/f_{1max} times. And each one of group n_3 will be scanned f_{3max}/f_{1max} times.

With scanning rate $f = 1/T = 100$ kHz, suppose:

$$n_1 = 100 \quad n_2 = 100 \quad \text{and} \quad n_3 = 56, \quad \frac{f_{2max}}{f_{1max}} = 10 \quad \text{and} \quad \frac{f_{3max}}{f_{1max}} = 100$$

We find:

$$f_{1max} < \frac{f}{2N}$$

Hence:

$$f_{1max} < 8 \text{ Hz}; f_{2max} < 80 \text{ Hz}; f_{3max} < 800 \text{ Hz}.$$

This introduces the need for anti-abrasing filters, which in practice reduce the spectra of the measuring signals, thus avoiding sampling frequencies that are too high. The effectiveness of a filter depends on its order. In most cases, Butterworth filters are a fair compromise between good amplitude selectivity and low phase distorsion.

9.1.1.4 Sampling rate

Sampling rate, which depends on the number of analogue inputs and the bandwidth of phenomena to be analysed, is expressed in channels per second or points per second. As it ranges from 10 points/s for slow systems, up to several hundred thousand points/s for very fast systems, it may be an important criterion of choice for system architecture. When sampling rate is higher than 10 000 points/s, the architecture commonly adopted includes a multiplexer, followed by a programmable gain instrumentation amplifier, a sample-and-hold, and an ADC. In order to limit errors due to parallel connection of several multiplexers, groups of 8–16 channels are made, including one multiplexer and one decoding logic. Although stacking several groups increases the capacity of the system, up to several hundred channels, the designer should remember that the leakage current of multiplexers increases with the number of channels, and generates significant errors at the inputs of the measuring amplifiers. Very often, the multiplexer must also ensure good insulation between transducers and acquisition system although *Reed-contact multiplexers* present real galvanic insulation which can withstand several hundred volts of common-mode voltage, they are limited to applications limited to rates of a few hundred channels/s. Their poor reliability, short life, cumbersome handling and large power consumption, explain the lack of interest in them for modern acquisition systems. Isolation amplifiers, with opto-coupler or transformer, can be a good solution to the problem, without significantly harming the system's sampling rate.

9.1.1.5 Conversion

Dual slope and quad-slope ADCs are extremely accurate converters as they can show a high serial-mode rejection. As their conversion time is longer than several tens of milliseconds, the Successive-approximation ADCs are clearly faster with conversion times, which depend on the number of bits and clock frequency, as short as a few microseconds. These converters are well suited for medium-to-fast acquisition systems although the input comparator makes them very sensitive to noise superimposed on signal. For very fast systems, parallel ADCs must be used.

ADCs are generally proceded by sample-and-holds. Choice of the latter is not only dictated by precision of sampling (offset, charge transfer, feedthrough, ...), but also by stability (for closed-loop circuits) and sampling frequency (which must be compatible with ADC resolution). Restoration of the analogue signal is done by a DAC. The designer should make his choice to meet the exact system requirement, bearing in mind that high-resolution converters are still very expensive. In order to preserve the DACs good resolution, it is often essential to take special care with layout. Some second-order errors, which are negligible on 12-bit conversion, may be significant on 16-bit conversion. This is true in the case of analogue reference zero error. By example, a current flowing from analogue signal ground to negative supply may vary from 0 to 2 mA, depending on the digital word applied, thus introducing a 66 μV offset ($\frac{1}{2}$ LSB on 16 bits) across a printed circuit board strap 15 cm long, 0.3 mm wide. Thorough understanding of the manufacturer's specifications is necessary, in order not to confuse the DAC's settling time, which is exceedingly small (for example 300 ns), the settling time of the current–voltage converter (several tens microseconds), often realized by using an op amp behind the DAC. The DAC also introduces an offset voltage, depending on the applied digital word. This offset voltage may seriously limit the operating range of the DAC, because of saturation of the current–voltage converter.

9.1.1.6 Transmission

When measurement points are geographically scattered as in a telemetry system, appropriate methods of transmission between these points and the processing unit must be chosen such as by adapted line with a suitable

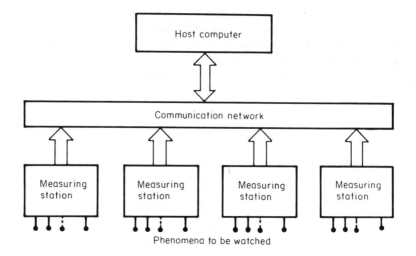

Figure 9.1 Organization of a distributed acquisition system

transmitter–receiver, or by modems and specialized telephone lines. Decentralization may lead to a distributed acquisition system, with several local subsystems called *measurement stations*. Each station acquires and processes the local data before transmitting them in digital form to a host computer through a communication network. So the host computer manages the whole system.

Data may be transmitted in parallel or serially, asynchronously or synchronously. Choice of a communication system is determined by number of stations and distance between stations, data flowrate (expressed in kbit/s), acceptable error, and contamination level due to environment. The host computer may be merely a microcomputer which should permit the development and updating of specialized software by using the potentialities of its monitor system (Figure 9.1).

The designer may consider the use of peripherals with large storage and display capacities such as CRT, VDU, disks, tapes, etc.

Connection of measurement stations and host computer are performed through 'standard' communication interfaces. They define material connections (characteristics of the electrical signals, wires, plugs) and communication procedures (initialization and transfer validation commands, designation of the employed frame, etc.). Among the most widely used 'standard' connections, we mention the RS 422 standard (which can transmit up to several Megabauds rate over 1.2 km), and the GPIB interface, which will be discussed elsewhere. In several process-control systems, coaxial or bifilar wound lines are used, at rates up to 1 Megabaud, at a distance of several kilometres.

9.1.2 Industrial Realization Examples

9.1.2.1 Data-acquisition module

From the last decade on, many electronic components manufacturers supply very versatile data-acquisition modules under hybrid structure. As an example we may mention the DT 5716 module, from Data Translation (Figure 9.2). This contains:

— One multiplexer, with either 8 differential channels, or 16 single-ended channels, at the user's choice.
— One instrumentation amplifier, followed by sample-and-hold. The gain is programmable, permitting a large analogue input range, from ± 5 mV to ± 10 V.
— A 16-bit ADC, with 50 μs conversion time. Three-state logic output makes it suitable for use with a data-processing system.
— A control logic, including input selection and sampling command.

The number of analogue inputs may be increased with an expansion module. For example, with the DT 48EX (Data Translation), we can obtain 64 single-ended or 32 differential inputs. This is a very accurate acquisition system ($\pm 0.0075\%$ of full scale, at unity gain), but relatively slow (350 μs acquisition time).

9.1.2.2 Acquisition system controlled by desk computer

A good example is the DAISY 12C system (Figure 9.3), designed by the Cell Neurobiology Laboratory (Centre National de Recherches Scientifiques, CNRS, France), and shown at the Physics-Mesucora Exhibition (Paris, 1983). The DAISY 12C system was originally designed for studies of the biological mechanism involved in the transmission of nerve impulses. The system may be used for acquisition, as well as for stimulation or control. The latter functions require an analogue output signal from DAC, which is frequently used in physiology to 'stimulate' biological structures (nerve membranes, among others).

(1) Description of DAISY 12C System

(a) *Acquisition module*
This includes:

(i) Differential input amplifiers, with gain adjustable from 1 to 50.
(ii) Data-acquisition module DAS-250 B (Datel), with one 16-input multiplexer, expandable to 32 inputs, one fast sample-and-hold which can acquire a voltage swing of 10 V in 200 ns with 0.1% accuracy, one successive-approximation 12-bit ADC with conversion time less than 2 μs, and the control logic which manages the sequencing of input sampling.

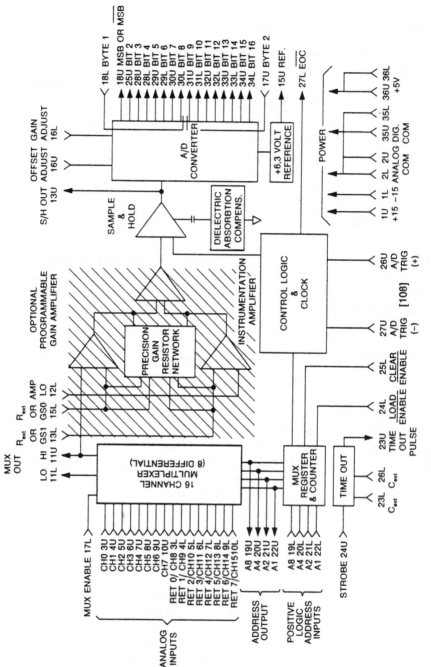

Figure 9.2 Example of data-acquisition module

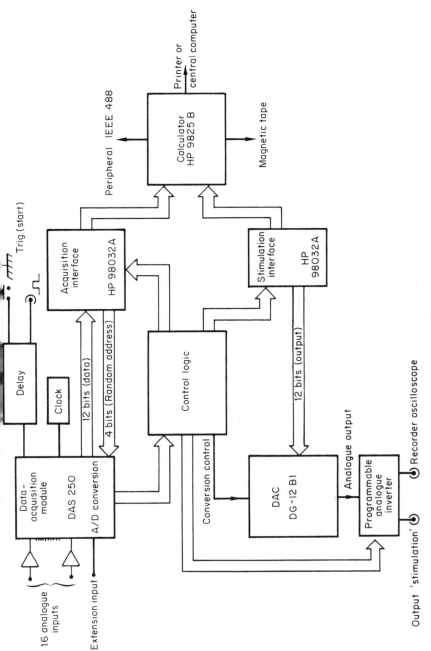

Figure 9.3 Functional block diagram of the DAISY 12C acquisition and stimulation system

(iii) An acquisition delay circuit, which enables acquisition only after a preset time. This is convenient when acquisition must be delayed with respect to stimulation.

(iv) A command logic synchronizes the end of conversion by the DAC (System Busy) and data-transfer request to interface (Peripheral Flag or PFLG).

(v) A clock with a 1 MHz crystal oscillator. The frequency is divided to obtain 250 kHz, and also three other submultiples by 2 of this frequency. Maximum sampling frequency is 250 kHz.

Moreover, the acquisition module has 3-state digital outputs which can be directly connected to the 4, 8, 12 or 16 bit bus of a majority of computers. The input channels may be addressed either sequentially or randomly. Address of the channel to be sampled is supplied on 4 bits by the computer.

(b) *Stimulation module*

This includes:

(i) Datel's DG-12B DAC. This 12-bit converter has been chosen for speed (settling time: 600 ns), and near absence of glitches, whose maximum amplitude is less than 2 LSB.

(ii) A programmable switch, which enables connection of the analogue output of the DAC, either to an oscilloscope or to a recorder, before the signal appears on the 'stimulation' output.

(iii) A command logic which synchronizes the DAC's conversion start signal with the handshake signals between interface and external peripherals (Peripheral Control or PCTL, Peripheral Flag or PFLG).

(c) *Acquisition and stimulation interfaces*

This acquisition and stimulation system is driven by a desk computer (Hewlett-Packard HP 9825 or 9826). The acquisition interface functions in DMA (Direct Memory Access) mode, while the stimulation interface functions in FRW (Fast Read Write) mode, as the calculator has only one DMA channel. Maximum transfer speed may reach 300 000 kilowords (16-bit) per second. The language used is HPL, a conversational language as easy to use as Basic.

(2) *Function of System*

(a) *Acquisition*

This may be set in automatic or triggered mode. In the first case, conversions are performed automatically on receipt of a transfer command from the computer. In the second mode, acquisitions take place only after an external trigger signal, and are allowed after a presettable delay, between 0.5 ms and 5 ms.

The sampling frequency is selected from the following: 250 kHz, 125 kHz, 62.5 kHz and 31.25 kHz. Lower frequencies may be obtained through modification of the acquisition program and external triggering. Sampling frequency of each channel and order of succession are programmable. Three operating modes are available:

(i) *Single channel*
The 'extension' input is used. Maximum sampling frequency is then 250 kHz: this allows maximum acquisition speed.

(ii) *Multi-channel, sequential mode*
The 16 channels are scanned automatically in fixed sequence 1–16. Maximum sampling frequency on each channel is then 250 000/16, i.e. 15.625 kHz. The number of channels may be reduced by shorting some of them with internal straps, and conversely the number of channels may be extended to 32 by using a second multiplexer (for example, Datel MX-1606), in conjunction with the 'extension' input, and the address sequencer outputs of the acquisition module.

(iii) *Multi-channel, random mode*
Any channel may be sampled randomly by transferring a 4-bit address to the acquisition module (4 bits are needed to address 16 channels). This address is computed by the calculator. This operating mode permits, first sampling at the maximum frequency (250 000 words per second) in case only one channel is needed, and on the other hand, scanning some channels more frequently than others, for example where phenomena must be watched more closely.

The acquisition chronogram is shown in Figure 9.4.

Acquisition is driven by the calculator. The transfer command (TFR) is

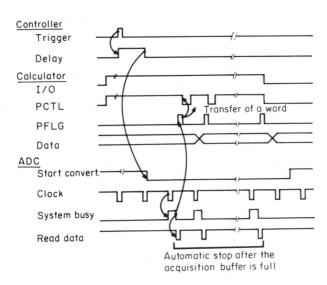

Figure 9.4 Chronogram of acquisition

translated on the control lines of the acquisition interface by:

— Level change from 'clear' to 'set', on peripheral control line (PCTL).
— Level change to 'high' on input/output line (I/O).

Transfer is then automatically performed, without processor control, until the acquisition buffer is full which provokes the return of PCTL and I/O lines to their initial level. Remember that acquisition is performed under DMA mode. Four output lines, corresponding to the four least significant bits of the acquisition interface, are used to address the active channel in the random mode. Sample size is controlled by the program from 1 to more than 16 000 words.

(b) *Stimulation*

This is done in 'Fast Write Read' mode, for the reasons already described, which limits transfer speed to the DAC to 80 000 words per second. Digital-to-analogue conversion is performed after a change of PCTL line on the stimulation interface from 'clear' to 'set' level, the I/O line being 'low'. This change is commanded by the computer (Binary Write Command (WTB), or Transfer Command (TFR)). The analogue signal is then directed to either outputs of the stimulation module, through the programmable switch.

(3) *Possible use of DAISY 12C System*

The main uses of this system may be:

— in the medical field fetching physiological data for EKG or ECG purposes, setting up dynamic and automatic monitoring systems;
— in the scientific field for the centralization of acquisition, process control, regulation, and so on.

The large capacity for processing digital signals possessed by the computer should not be forgotten. Overall performance is limited by the HP 9825 computer, which may be replaced by the HP 9826 (Figure 9.5). In addition to the HPL language, compatible with programs written on HP 9825, the HP 9826 supports Pascal, which notably speeds up data processing. It has two DMA channels, and can drive the acquisition and stimulation modules simultaneously in this mode, thus enhancing the output rate of the analogue signals. This is valuable for accelerating the response of a process-control loop.

9.1.3 Measurement and Control System

Besides the data-acquisition modules including programmable gain amplifiers, differential inputs, multiplexers, etc., many other intelligent measurement and control subsystems are possible. These include signal conditioning,

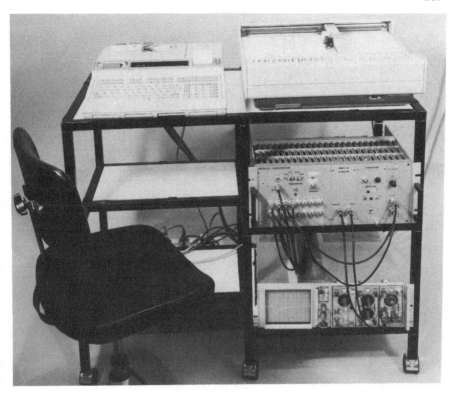

Figure 9.5 General view of Daisy 12C system

isolation, microcomputer-based linearization, communication with host computer, a.c./d.c. conversion, d.c./d.c. conversion, digital inputs, latched outputs, etc. These are manufactured in the form of discrete-assembly system, which may help to solve a broad range of industrial measurement and control problems. Modular design allows expansion within the system with the flexibility to accept a variety of analogue and digital input and output functions.

They are optimized for high performance measurements. Analogue input handling capability offers reliable operation in noisy industrial environments. The high performance is realized by high-quality signal conditioning, high input to output isolation, high common-mode noise rejection, efficacious filtering, low drift amplification, high-resolution analogue-to-digitals conversion.

Figure 9.6 shows a functional block diagram of a standard measurement and control system.

(1) Many different input sources can be applied to the analogue inputs from strain gauges, RTD's, thermocouples, pressure transducers, etc.

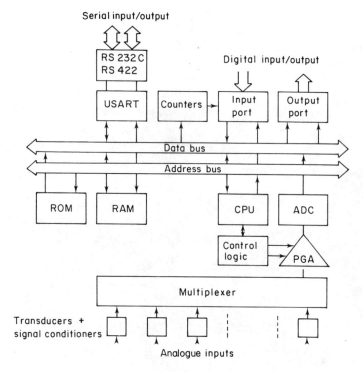

Figure 9.6 Measurement and control system functional block diagram

Signal-conditioning modules provide bridge excitation, isolation, pre-amplification and filtering. These modules are followed by a multiplexer.

(2) The gain select of the programmable-gain amplifier (PAG) is controlled by digital logic. It is very useful for amplifying the preamplified input signal to the full-scale range of the following ADC.

(3) The analogue-to-digital conversion may be performed by a flash converter, which has a resolution depending upon the desired number of conversions per second.

(4) The computer and its operating system may use optional ROM and RAM. A DIP-switch option permits programs to be run from either ROM or RAM.

(5) The system may be designed to operate in stand-alone applications and may communicate with any host computer. Communication by two serial ports is possible. A dual programmable communication interface is used to receive and transmit data at selectable baud rates; this is the Universal Synchronous/Asynchronous Receiver Transmitter (USART).

The local and remote ports may communicate in RS 422 or RS 232C.

The RS 232C interface is realized by a 25-line connection. It is fitted for

Table 9.1 Functional definition of the usual pins of the RS 232C interface.

Pin	Symbol	Transfer direction	Function
1	FG	↔	Frame ground
2	TD	→	Transmit data
3	RD	←	Receive data
4	RTS	→	Request to send
5	CTS	←	Clear to send
6	DSR	←	Data set ready
7	SG	↔	Signal ground
8	DCD	←	Data carrier detect
20	DTR	→	Data terminal ready
22	RI	←	Ring indicator

Transfer direction is indicated by:
→ from computer to peripheral
← from peripheral to computer

both synchronous and asynchronous communications. Serial data transmission is performed on a unidirectional transfer line.

The maximum connection length is 50 ft (15.2 m) with a flowrate of 20 kbaud. The logic 0 is defined by a voltage of +3 V, the logic 1 is defined by a voltage of −3 V. Therefore the 'start' bit preceding a data asynchronous transmission is indicated by 0, while the 'stop' bit is indicated by 1. This standard is very similar to the CCITT/V.24 standard (CCITT: Comité Consultatif International de Télégraphie et Téléphonie), which determines more particularly the interface between a modem and a data transmission terminal.

The RS 422 interface has been used since 1976. The maximum connection length is 4000 ft (1.2 km) with a flowrate attaining 10 Mbaud. The signal voltage is defined by the difference between voltages of the two transmission lines, which is completely independent of ground.

(6) Digital input/output ports offer interface to contact closures, TTL levels or high-level a.c. and d.c. voltages. Counters provide for pulse accumulators and frequency inputs.

9.1.4 Measurement Stations

This term is used to describe systems for the centralized collection of measurements, local supervision and industrial control instruments. They have been introduced on to the market since the appearance of digital voltmeters. Initially used merely to supervise and record process parameters, measurement stations have evolved in parallel with programmable digital techniques. The number of functions, flexibility of use and versatility have increased. They are now used in many fields.

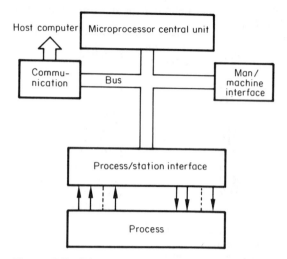

Figure 9.7 Diagram of a standard measurement station

They generally include four main functions:

— Interfacing between the process and the measurement station, which is in charge of the acquisition of analogue quantities, signal conditioning, electrical isolation, conversion and generation of control signals.
— Control of the station, and measurement processing, performed by the microprocessor central unit.
— Man–machine interfacing which enables better and easier use of the station's possibilities. It consists of keyboard, display, printer, recorders, etc.
— Eventually, a linkage with a host-computer, to build-up an acquisition system distributed over several other stations.

Obviously, the control and processing unit is the critical part of a station. It contains one or several microprocessors associated with PROMs containing the manufacturer's software, and RAMs containing the user's programs, unprocessed data, calculation results and processed data. The cost of a station varies notably with the capacity of the central processor, and volume of software supplied.

Simple systems have a central unit with one 8-bit processor, 32-kbyte PROM and 32-kbyte RAM. More powwerful systems use multiprocessor central units. Each processor has its own memories and is devoted to one particular function. For example, the first one acquires data, the second controls peripherals, the third processes the acquired data, etc.

9.2 DIGITAL PROCESSING OF DATA

Digital processing of signals deals with a variety of operations including spectral analysis, correlation, filtering, transcoding, modulation, detection, estimation and smoothing) which should be performed with a digital processor in order to extract useful information from the received signal. Digital treatment has arisen from the development of fast computing algorithms, which permit real-time processing in many applications.

Digital filtering is one of the most important digital processing techniques. Its scope extends to all activities involving acquisition and data processing.

Until recently it has been considered as an expensive technique to be used only for those particular problems requiring large calculation and processing volumes requiring a computer. The increasing emergence of microprocessor-associated peripherals and microprogrammed modules now permits the setting up of digital filters which are self-operated, economic and of increasingly good performance. This trend leads us to predict ever wide use of this technique in the near future.

The main benefit of digital filtering is its great flexibility for frequency adjustment. For example, the characteristic frequency is modified only by changing the sampling frequency, without modifying software or hardware. Moreover, it allows realization of linear dephasing filters, abrupt cut-off filters, and filters with very stable characteristics.

Disregarding cost a major drawback is certainly the computing time, which is often too long, in particular for signals above a few tens of kilohertz. At the present, self-contained circuits have been developed, at decreasing price, permitting prediction of a notable enhancement of processing speed in the coming years. As examples some recent products can be cited.

9.2.1 Digital Multiplier-accumulators (MAC)

Now that low-cost multiplier accumulators (MAC) are readily available, it is possible to implement many digital signal-processing (DSP) applications. These new devices make DSP techniques possible in the electronic area. In particular, it simplifies digital filter design.

There are many synthesis methods and many designs. Within the scope of this work, we present as an example the synthesis of *finite impulse response* or FIR filters which show numerous advantages. As the phase response can be made linear it is important in numerous fields where this interesting property is useful, especially data transmission. The filter has no pole so that it is thus always stable Lastly it permits good accuracy due to the neglect of the problems of quantization and rounding.

The filter equation can be written

$$Y_n = \sum_{k=0}^{n-1} h_k X_{n-k}, \qquad (9.1)$$

where Y_n is the digital output of filter at time n
X_{n-k} are the digital inputs
h_k are coefficients corresponding to the impulse response of filter.

To understand the function of the filter described by this equation, recall that the convolution of two signals in the time domain is equivalent to multiplication in the frequency domain. The Z-transform of Y_n is:

$$Y(Z) = H(z)X(Z) = \left[\sum_{k=0}^{n-1} h_k Z^{-k}\right] X(Z), \quad (9.2)$$

In discrete harmonic analysis this is equivalent to

$$Y(f) = \left[\sum_{k=0}^{n-1} h_k e^{-2\pi i f k T}\right] X(f) = H(f)X(f), \quad (9.3)$$

where $H(f) = \sum_{k=0}^{n-1} h_k e^{-2\pi i f k T}$ is the numerical Fourier transform of h_k. It is a periodic function of f, with period $1/T$. Reciprocally, h_k are the coefficients of the Fourier series development of $X(f)$.

To synthesize the filter the number and values of the coefficients h_k should be determined. There are several available techniques, which cannot be presented within the scope of this work. The reader is referred to books cited in the references. The implementation of the filter is readily obtained from its coefficients. Equation (9.1), relative to a 5th-order filter, may be represented as follows (Figure 9.8).

The pure delays Z^1 may be realized by a microprocessor or a shift register.

The multiplier and accumulator, which are the 'nerve centre' of the filter, are included in a single module ADSP 1110 (Analog Devices). Note that each digital output value is the sum of products of the five preceding entries by five coefficients. Hence the output signal Y_6 is

$$Y_6 = X_6 h_0 + X_5 h_1 + X_4 h_2 + X_3 h_3 + X_2 h_4. \quad (9.4)$$

A single multiplier may be used, with intermediate products memorized in the accumulator. The frequency response of the filter may be increased

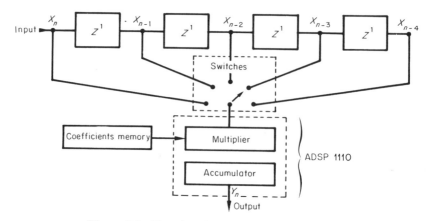

Figure 9.8 Functional diagram of a digital filter

by using more multipliers. Thanks to the symmetry of the coefficients h_k of the finite impulse response filter, one can write its equation in the following form:

$$Y_n = h_0[X_n + X_{n-4}] + h_1[X_{n-1} + X_{n-3}] + h_2 X_{n-2}. \quad (9.5)$$

The number of multiplications is then divided by two. This leads one to think that with one extra accumulator and control logic it may be possible to double the filter bandwidth.

The recently introduced single port 16×16 MACs may be used for this kind of filter design. These new MACs have a very interesting role to play in DSP. By contrast with classical filtering techniques, they are of great use in performing matrix multiplications, correlation, fast Fourier transform (FFT), spectral estimation, etc. The most useful application is probably the FFT computation.

FFT is a generic term for a collection of algorithms for efficiently calculating the discrete Fourier transform (DFT), and the inverse discrete Fourier transform (IDFT). Recall that the DFT of a finite length sequence is defined as

$$X(k) = \sum_{n=0}^{N-1} x(n) e^{-j2\pi kn/N} \quad \text{for } 0 \le k \le N-1.$$

The IDFT of $X(k)$ is

$$x(n) = \frac{1}{N} \sum_{k=0}^{N-1} X(k) e^{j2\pi kn/N} \quad \text{for } 0 \le n \le N-1.$$

Note that the above expressions for DFT and IDFT differ only in the sign of the exponent of the phasor $\exp(j2\pi kn/N)$ and in a scale factor $1/N$. Hence computational procedures for the DFT can be easily modified to compute the IDFT. The most straightforward method of calculating an N-point discrete Fourier transform using the definition directly requires N^2 multiplications corresponding to N multiplications for each of the N frequency points $X_0, X_1, \ldots, X_{n-1}$. However, since the amount of computation (and thus the computation time) is proportional to N^2, the number of computations required for the DFT by such a direct method becomes very large for larger values of N. The FFT subroutine is an algorithm which reduces the number of computations required for the DFT from N^2 to $N \log_2 N$, where N is a power of 2. Because of the periodic and symmetry properties of the quantities $\exp(-j2\pi kn/N)$, many of the multiplications and additions can be eliminated. In fact, the amount of numerical computation can be considerably reduced.

Example Assume that $N = 2^{10} = 1024$. Normally we should have $N^2 = 1\,048\,576$ operations. The FFT algorithm may allow performance of only $N \log_2 N = 10\,240$ multiplications and additions.

The details of many FFT algorithms can be found in references. The basic arithmetic operation in an FFT algorithm is a multiply-accumulate operation, which can be easily and efficiently realized by a single-port MAC.

The fundamental principle of the FFT is to decompose the computation of the DFT of a sequence $x_0, x_1, \ldots, x_{n-1}$ of length N into successively smaller DFT. In this process, the FFT algorithm 'shuffles' the input data $x_0, x_1, \ldots, x_{n-1}$ and combines the data by adding sums pairwise, then adding pairs of sums pairwise, and so on. This merging procedure involves the use of a quarter-length cosine table, and repeatedly involves the so-called *butterfly operation*. To complete a butterfly, a number of multiplications, additions and subtractions have to be performed. By properly sequencing the butterflies we can obtain an FFT. For example, a 1024-point FFT contains about 5000 butterflies. A single-port MAC is of great use in implementing butterflies. The ADSP 1110 (Analog Device) requires 100 ns per cycle. Assume that for each point ten cycles may be required to compute the real component and ten other cycles for the imaginary component, the total time required is 2.0 ms. A 1024-point FFT requiring about 5000 butterflies takes 10 ms.

9.2.2 Digital Signal Processor

This new digital signal-processing family is providing the design engineer with a new approach to a very large variety of complicated applications. Digital signal processors are single-chip microcomputers which combine the flexibility of a high-speed controller with the numerical capability of an array processor. They are capable of executing several million instructions per second. This high throughput is the result of the efficient and easily programmed instruction set and of the highly pipelined architecture.

Before looking at digital signal processing, it is useful to clarify the distinction between signal processing and digital processing. Most usual microprocessors are designed for data processing, not for high-speed complex signal processing. The 8080/8085 microprocessor system can operate as a signal processor at frequencies to a few hundred hertz, and requires multiple chips with a separate analogue-to-digital converter and input/output system. Thus a general-purpose microprocessor is not well-suited for signal processing applications. A different processor architecture is required to implement signal-processing algorithms.

Until recently digital signal processing was only cost-effective when it could be applied in large systems. Now, however with VLSI techniques, low-cost processors are available and many opportunities exist for the application of DSP techniques. The potential applications may be found in any measuring system where signals contain useful information generated by measuring transducers. Generally the objective in processing the signal may be to prepare it for digital transmission or storage. The signal may be processed to eliminate the distortion introduced by transducers, or by analogue signal conditioners. Certain processors are capable of providing the multiple functions required for a single application. Thus they may allow a robot to recognize speech and to perform mechanical operations through digital servo-loop computations. In the field of instrumentation, the architecture of these processors may be especially suited to the implementation of the basic DSP algorithms for recursive and non-recursive linear filtering, discrete Fourier transformations (DFT) and spectral analysis. In the following paragraph the basic architecture of modern DSP and some usual applications in measuring systems will be described.

The basic architecture of a modern DSP is shown in Figure 9.9. It contains four main elements:

— An *arithmetic logic unit* (ALU) which adds, subtracts, and performs logical operations. Generally it operates on 32-bit data words.
— An *accumulator* which stores the output from the ALU and is also often an input to the ALU. It is divided into two words (a low-order word and a high-order word), which will be stored in data memory.
— A *multiplier* which can perform a 16 × 16 bit, 2s complement multiplication. It consists of three devices: the *T*-register, the *P*-register and the multiplier array. The *T*-register temporarily stores the multiplicand; the *P*-register stores the product. This multiplier may allow performance of some common operations (for example: convolution, correlation, filtering, etc.) at the rate of 2.5 million samples per second.
— *Shifters*: A barrel shifter for shifting data from the data RAM into the ALU, and a parallel shifter for shifting the accumulator into the data RAM are the two shifters available for manipulating.

As shown in Figure 9.9, program memory and data memory lie in two separate spaces so that transfers between program and data spaces are allowed, thereby increasing the flexibility and speed of the device by this architecture.

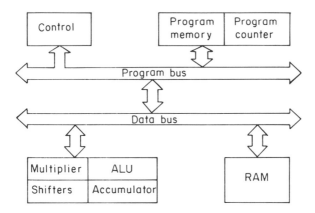

Figure 9.9 Architecture of modern DSP

The instruction set of these DSPs consists primarily of single-cycle single-word instructions allowing execution rates of up to several million instructions per second. It contains a full set of branch instructions which permit the bit-manipulation and bit-test capability needed for high-speed control operations.

Example of industrial realization
TMS 32010 16/32 bit Digital Signal Processor (Texas Instruments) Main features:
— 200 ns instruction cycle
— 288 on-chip data RAM
— 3 kbyte on-chip program ROM – TMS 320M10
— external memory expansion to a total of 8 bytes at full speed
— 16-bit instruction/data word
— 32-bit ALU/Accumulator
— 16 × 16-bit multiply in 200 ns
— 0 to 15-bit barrel shifter
— 8 input and 8 output channels
— 16-bit bidirectional data bus with 40 Megabits per second transfer rate.

Since highly efficient computation of the DFT is possible, it is natural that many uses have been found for the DFT. Figure 9.10 indicates a scheme realizing a discrete convolution using the FFT where $x(n)$ and $h(n)$ are convolved to produce $y(n)$. To benefit the great efficiency of the FFT, it is more advantageous to compute $X(k)$ and $H(k)$ multiply $X(k)$ and $H(k)$ together, and compute $y(n)$ using the inverse Fast Fourier Transform (IFFT), rather than to compute $y(n)$ directly by discrete convolution.

This technique may also be used to determine correlation functions since correlations can be computed by time reversal of one of the sequences before convolution.

Another important area of digital signal processing is spectrum analysis. Different methods may be used. The classical approach is to sweep the input signal through a narrow bandpass filter and observe the filter response as a function of the frequency. Figure 9.11 shows the implementation of a spectrum analyser. The input signal is first shaped by a low-pass filter to avoid aliassing from the overlapping spectral components after multiplication. The filtered signal is then multiplied by the sweeping local oscillator to

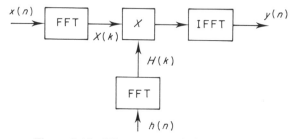

Figure 9.10   Discrete convolution using FFT

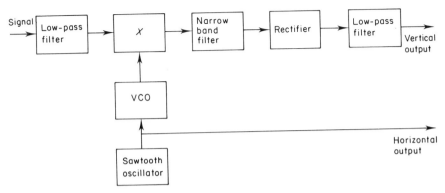

Figure 9.11 Block diagram of a classic spectrum analyser

generate upper and lower sidebands. Only the upper sideband is of interest, however, as it is swept across the band-pass filter and the signal energy extracted. The sweep output provides a horizontal sweep voltage for an X–Y display. It is natural that in the digital implementation there must also be a sample-and-hold and an ADC at the input, and a DAC and a reconstruction filter at the output. The functions shown on Figure 9.11 are realized by certain analogue signal processors (such as the 2920 from Intel) which generally offer a narrow input bandwidth (<4 kHz).

Another spectrum analysis method relies on the computation of estimates of the Fourier transform or the power spectrum of the signal. The analogue signal must be low-pass filtered and digitized. The major concern in this application is the need to truncate the input signal to a finite length for the FFT computation. This can be represented as a *windowing* operation which gives a finite-length sequence from $x(n)$ by

$$\begin{cases} y(n) = w(n)x(n) & \text{for } 0 \leq n \leq N-1 \\ y(n) = 0 & \text{otherwise.} \end{cases}$$

The Fourier transform of $y(n)$ is

$$Y(e^{j\omega T}) = \frac{1}{2\pi} \int_{-\pi}^{+\pi} X(e^{j\theta T}) W(e^{j(\omega - \theta)T}) \, d\theta.$$

where $X(e^{j\omega T})$ is the Fourier transform of the input signal, and $W(e^{j\omega T})$ is the Fourier transform of the window. The FFT is the most complex of the spectrum analysis functions. Some modern DSP's can perform a 64 complex

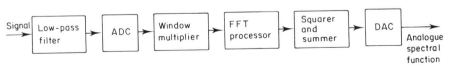

Figure 9.12 FFT-based spectrum analysis

points FFT in only 580 μs; much larger transforms can be performed by using off-chip RAM.

The FFT-based computation of a spectrum can also be used in speech, radar and sonar signal processing.

9.3 DIGITAL INTERFACE FOR PROGRAMMABLE INSTRUMENTATION

9.3.1 Automatic Measurement System

When the measurement system is intended to effect a series of well-specified operations and to repeat them a large number of times (manufacturing line tests, for example), it is often of interest to automate the functioning of the system.

An automatic system can be designed by connecting different devices in a well-defined order, so as to execute the diverse operations without human supervision. The structure of the system is obviously very specialized and complex, and does not support any modification, because each system has special connections, whether mechanical (connectors) or electrical and functional (signal amplitude, coding, etc.).

Electronics and data processing allow an economic solution, which is also flexible, due to the standardization of digital interfaces for programmable instrumentation. This is the subject of the international standard IEEE 488, released in April 1975. The standard prescribes mechanical specifications (connector type, pin assignment, etc.) and electrical specifications for interface (logical–electrical state relationships, characteristics of the controlling devices, receivers and interconnecting cable, etc.). Direct interconnection of any devices, with a single standardized cable, is thus possible, with the only requirement that they include an interface circuit meeting the IEEE standard. Figure 9.13 shows a simplified structure of the system, where the user is replaced by a device called the *controller,* connected to all others by a bus line.

The controller is in charge of management of all data transfers. Data transmission and reception are the two kinds of function distinguished. The

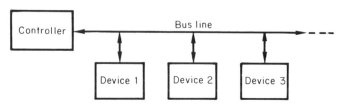

Figure 9.13 Simplified structure of an automatic measurement system

controller must determine which devices have to participate in data transfer, and which function each one has to perform. Each device must thus have a proper address.

Three kinds of functions can be discerned in any automatic measurement system: selection of a device, data transfer, and data transfer management.

The notion of transfer management is very important here, as a 'talker', (which could be the controller itself when transmitting addresses, or the selected device when transmitting data) cannot perform the next transfer until it is certain that all concerned devices have received the former data without error. When transfer is synchronous, its rate is then defined and determined by the maximum response time of devices whereas in asynchronous transfer the procedure is such as to cater for the speed of the slowest device participating in the transfer. All messages exchanged are routed on the bus. Each kind of message can only be transmitted on particular groups of lines:

— address lines;
— data lines;
— data-transfer management lines.

9.3.2 The General-purpose Interface Bus System (GPIB)

9.3.2.1 Description

The GPIB system is made up by functional, electrical and mechanical elements of an interface complying to the IEEE 488 standard. It is at first a connecting cable with 24 wires, where 16 are divided into three groups:

— *Data transfer lines*: 8 lines are used for bidirectional transfer of coded messages. These can be either commands or data from devices and are organized as sequences of 8-bit words.
— *Data transfer control lines*: 3 lines are used to manage the asynchronous transfer of information on the preceding 8 lines, between one 'talker' and one or more 'listener' devices (Listener devices are those which only receive information in the form of either commands or data).
— *General management lines*: 5 lines. These indicate the nature of information on the previous lines (either address or data), and execute some special functions which permit expansion of the capabilities of the interface system.

Figure 9.14 shows the assignment of the IEEE 488 bus lines, and Table 9.2 shows the names of the connector pins. In order to avoid cross-talk between the main control wires (DAV, NRFD, NDAC, IFC, ATN, EOI, REN and SRQ), each of them is individually shielded. Table 9.3 gives the electrical and logical levels of talkers and listeners. Notice that talkers have open-collector outputs.

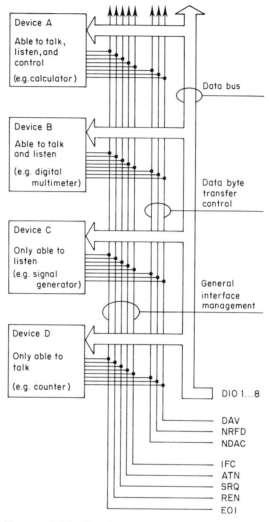

Figure 9.14 Interface capabilities and bus structure

9.3.2.2 Bus management

Bus management includes two parts:

(1) Data transfer management (lines DAV, NRFD and NDAC)

— The DAV line (Data Valid) is a validation line, activated by the talker to inform the listeners that data are available on the bus.
— The NRFD line (Not Ready For Data) is used to inform the talker that the listeners are ready to receive data.

Table 9.2 Connector contact assignments

Pin	IEEE Standard Designations
1	DIO 1
2	DIO 2
3	DIO 3
4	DIO 4
5	EOI
6	DAV
7	NRFD
8	NDAC
9	IFC
10	SRQ
11	ATN
12	SHIELD
13	DIO 5
14	DIO 6
15	DIO 7
16	DIO 8
17	REN
18	GND 6
19	GND 7
20	GND 8
21	GND 9
22	GND 10
23	GND 11
24	LOGIC GND

— The NDAC line (Not Data Accepted) serves to indicate that the listeners have finished reading data.

Management proceeds as follows. The controller addresses talker and listeners which bring line NRFD to 1, informing the talker that they are

Table 9.3 Logical and electrical state relationships

GPIB input	GPIB output	Logical state	Current or voltage Minimum	Maximum
V_{low}		1	−0.6 V	+0.8 V
V_{high}		0	+2.0 V	+5.5 V
	V_{low}	1	0.0 V	+0.4 V
	V_{high}	0	−2.4 V	+5.0 V
I_{low}		1		−1.6 mA
I_{high}		0		+50 μA
	I_{low}	1	−48 mA	
	I_{high}	0		−5.2 mA

ready to receive data. The talker then sends data on the bus, and brings line DAV to 0, indicating to the listeners that data are available on the bus. The listeners store these data in memory, then bring line NDAC to 1, announcing end of reading. The talker then brings line DAV to 1. The listeners acknowledge by bringing lines NDAC to 0, and NRFD to 1, to indicate that they are again ready to accept further data.

(2) General management of bus (lines IFC, ATN, SQR, REN, EOI)

— Line IFC (Interface Clear) permits the controller to initialize all connected instruments in a well-defined state.
— Line ATN (Attention) enables the controller to indicate that instructions and addresses (ATN at 1) or data (ATN at 0) are available on the bus. When ATN is at 1, only talker and addressed listeners are concerned.
— Line SQR (Service Request) enables any instrument to ask for the controller.
— Line REN (Remote Enable) puts all connected instruments under bus control.
— Line EOI (End Or Identify) is activated by the talker. When ATN is at 0, it indicates that data on bus are the last ones. When ATN is at 1, it initializes a polling sequence.

(3) Coding

Table 9.4 summarizes the digital command codes and addresses of talkers and listeners. How can connected devices distinguish between the information on these 8 lines?

The logical state of ATN permits this when ATN is at 1, the information on the bus are addresses or command whilst they are data when ATN is at "0".

Data may be digits or alphanumerical characters coded in ASCII.

An address is assigned to each device. There are two kinds of addresses/commands:

— Primary addresses/commands amount to 64 divided between 32 for talkers and 32 for listeners.
— Secondary addresses, amounting to 32 in number, are associated with each primary address/command. For example, the primary address selectes the device, and the secondary address selects a function (as in a multimeter) or a measuring range.

Table 9.4 Message code

Least significant bits					Most significant bits							
				Binary codes	b_7	0	0	0	0	1	1	1	1
					b_6	0	0	1	1	0	0	1	1
					b_5	0	1	0	1	0	1	0	1
b_4	b_3	b_2	b_1	Hex	0	1	2	3	4	5	6	7
0	0	0	0	0	—	—	00	16	00	16	00	16
0	0	0	1	1	GTL	LLO	01	17	01	17	01	17
0	0	1	0	2	—	—	02	18	02	18	02	18
0	0	1	1	3	—	—	03	19	03	19	03	19
0	1	0	0	4	SDC	DCL	04	20	04	20	04	20
0	1	0	1	5	PPC	PPU	05	21	05	21	05	21
0	1	1	0	6	—	—	06	22	06	22	06	22
0	1	1	1	7	—	—	07	23	07	23	07	23
1	0	0	0	8	GET	SPE	08	24	08	24	08	24
1	0	0	1	9	TCT	SPD	09	25	09	25	09	25
1	0	1	0	A	—	—	10	26	10	26	10	26
1	0	1	1	B	—	—	11	27	11	27	11	27
1	1	0	0	C	—	—	12	28	12	28	12	28
1	1	0	1	D	—	—	13	29	13	29	13	29
1	1	1	0	E	—	—	14	30	14	30	14	30
1	1	1	1	F	—	—	15	UNL	15	UNT	15	31
					ACG	UCG	LAG		TAG		SCG	

Primary command group: PCG
  ACG: Addressed command group
  UCG: universal command group
  LAG: Listen Address group
  TAG: Talk Address group
Secondary command group: SCG

9.3.2.4 Simple examples

Example 1 Figure 9.15 shows a talker (for example a voltmeter) connected to a listener (for example a printer).

Figure 9.15   A talker is connected to a listener

Example 2   The controller may be a microprocessor which controls devices and processes data. These devices may be a voltmeter, a printer and a tape punch. The microprocessor adjusts the voltmeter, then connects it to the printer and the punch, so as to record results (Figure 9.16).

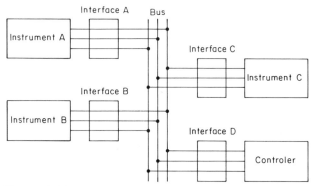

Figure 9.16   Instruments A, B, C are connected to a controller

Example 3   In evolved structures, groups of devices are connected through the bus. The system controller may allocate tasks to each group (Figure 9.17). It may also transfer control to a group controller, and take it back only when this group has finished its task.

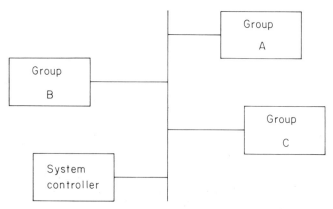

Figure 9.17   Architecture of a complex system: each group includes several pieces of apparatus

9.4 CHOOSING BETWEEN ANALOGUE AND DIGITAL TECHNIQUES FOR DESIGN OF A MEASUREMENT SYSTEM

We have witnessed during the last decade a growing trend towards digitization of measurement, information processing and control. In order to compare analogue to digital techniques, we shall exclude *a priori* those analogue techniques related to power systems, which are generally superior for these purposes.

9.4.1 Advantages and Drawbacks of Analogue Techniques

9.4.1.1 Drawbacks

(1) Noise

The two sources of noise which can be distinguished are internal noise due to the system's components, and externally induced noise. The latter is much stronger than the former. It is quite difficult to eliminate although shielding and grounding techniques can minimize it. However, the user must give special attention to the type noise superimposed on useful signals from sensors either at the measurement location or through pick up in the industrial environment. This class of noise often shows larger amplitudes than the useful signals, making filtering very difficult. Synchronous detection and correlation techniques provide solutions, which still remain delicate and expensive. The continuous nature of the analogue method, gives the quantitative aspect of a measured signal prime importance, so that the problems of noise are especially acute, as it interferes with the uncertainty on the value of the measured quantity. Noise influence is eliminated in digital systems, as here the quantity is worked out in a rather qualitative way, due to the effect of quantization.

(2) Drift

Drift appears as a time-dependent displacement of the measurement origin. It may lie at two levels:

At the level of the parameters of the system's components. Their variations induce changes in linear operating points, thus giving rise to parasitic offset voltages.

At the level of the manual adjusting devices (in zero-detecting methods, balance, etc.). Settings may be modified by mechanical vibrations, which induce very small displacements of potentiometers, or also by thermal phenomena, which generate dilatation of the adjusting elements.

Parasitic drift is a worrying matter in designing an analogue system. It increases costs, often quite significantly. In fact, its influence has to be minimized either by using extra compensation devices (drift compensator, zero automatic setting, etc.), by choosing stable components (metal or carbon film resistors, etc.). or by design (integrated differential amplifier, etc.). The drift phenomenon often makes zero adjustment useless in analogue instrumentation. On the other hand, in digital techniques, zero is only defined to within a quantum (±1 LSB).

(3) Difficulty in Memorizing the Measured Quantity

This is a big failing of analogue techniques, as to hold a measured quantity for a fairly long time, the only method is to use a capacitor, the defects of which are well-known. Magnetic recorders can perform very well as memories, in spite of limitations in noise and bandwidth. As access to information is very slow because of the sequential nature this prevents their use as temporary storage in analogue systems.

(4) Risks of Instability

Parasitic couplings can generate unforeseeable effects, leading to instability, which appears as oscillations of the system, making it useless. Instability may also result from variations of some parameters with temperature, increasing the system's gain. Recall that instability appears as soon as closed-loop gain is higher than 1, with $-180°$ phase lag.

(5) High Power Consumption

Analogue systems generally exhibit higher power consumption than digital systems, as the operating points of active components must be maintained in linear zones.

9.4.1.2 Advantages

(1) Predominance of Analogue Techniques in Current Transducers Manufacture

The transducer domain is still governed by analogue techniques. A majority of transducers deliver analogue quantities, such as voltages or currents. Only a few quantities, like angular or linear displacements, can be directly converted into digits. Digital speed transducers, which is derived from angular sensors exist, however digital information is obtained only after a time-consuming process, which is not the case with a tachometric generator.

(2) Simplicity in Designing

Provision of a model in a form familiar to design engineers is made easy because linearization is possible around the operating point. This is not the case for digital systems, where modelling is difficult or impossible because of their discrete nature.

Some analogue functions present unequalled performance at unchallenged cost:

— Analogue filters remain essential at high frequencies, because digital techniques suffer from frequency limitation due to speed of computation.
— Analogue function generators give unsurpassed results, as they maintain continuity of signal derivatives. In particular, sine-wave generators are much easier to build by analogue methods than digitally, although amplitude and frequency stabilities remain difficult to maintain.

(3) Ease of Realization

At the realization stage, there always remains the possibility of adjusting some variable elements to achieve optimal functioning. This does not apply to digital systems, which need time-consuming handling, as often the entire program may require modification, thus increasing the time required for development and validation of the software. Finally, glitch phenomena, which are found in logic systems, do not exist in analogue ones, because all signals are real-time with their instantaneous values.

9.4.2 Advantages and Drawbacks of Digital Techniques

9.4.2.1 Drawbacks

(1) Effects of Quantization and Sampling

Digitization of signals requires sampling followed by quantization. Quantization of a signal adds a white, uncorrelated noise with r.m.s. value equal to $q^2/12$. To reduce this, one obvious way is to reduce the quantum q in other words to increase the number of bits. Sampling turns an analogue signal into a discrete-time one. Shannon's theorem gives theoretical limits for the sampling frequency. This frequency is also limited by the quantization time. To put this another way, the more the accuracy which is required, the shorter and the more numerous will be the quantization intervals required leading to longer conversion times and more expensive devices.

(2) Impossibility to Access to Intermediate Values

In an analogue measurement system, observing the intermediate quantities at different locations within the chain is trivial. This is essential at the realization stage, and if a fault arises it is necessary to look for the defective element.

In digital systems, this is impossible for two reasons:

— The nature of digital information makes it difficult to ascertain, for example, capacity overflow (when the word is longer than the standard one). The information obtained makes no sense unless precautions are

taken. On the other hand, in analogue systems, the signal cannot grow above the saturation level imposed by the system.
— Use of more and more complex integrated circuits does not simplify the search for the cause of errors. This is particularly troublesome in programmed systems.

9.4.2.2 Advantages

(1) Insensitivity to Noise and Drift

All electronic devices inside the digital chain receive and process binary information. They test whether the voltage received is at level 'high' or 'low'. The amplitude of signal no longer interferes. As '0' and '1' levels may vary in quite large intervals, drift becomes unimportant here, because it always stays in a tolerable interval. The same applies to noise, as long as its amplitude remain insufficient to induce level change. There are large induced noise signals, which may provoke untimely switchings. The interference very often fed in by the mains power supply, which is perturbed by power-switching devices. The errors induced may have serious consequences, because of bit weighing. In programmed systems, this kind of error can be avoided, by assigning to each variable a maximum range of variations. Other techniques involving redundancy or error detection and correction may also be used.

(2) Ease of Transmission

Transmitting digital data is very easy, as logical levels can be readily restored, as long as signal amplitude permits distinction between high and low levels. It should be noticed that digital data is regenerated when the complexity of a system increases, whereas analogue data is degraded. This advantage is the reason why analogue-to-digital conversion is more and more performed in the immediate proximity of sensors located remotely from the measuring instruments.

(3) Good Galvanic Insulation

Galvanic insulation of systems controlling power devices is generally provided by transformers in analogue techniques. In digital ones, insulation may be better and easier, because it can be set up with optocouplers. Notice that reconversion to analogue signals does not cause any problem. If the DAC delivers step-shaped voltages, this shape is not always a constraint, as the subsequent devices (driven by this voltage) show generally narrow bandwidths, and filter the signals effectively.

(4) Ease in Digital Processing

The discrete nature of digital information has permitted quick development of mass produced electronic components for binary processing. A large number of complex functions may also be included. The microprocessor has become a component like other logic components. This privileged situation, which has grown with the decrease in prices during the last decade, is a very appreciable advantage for the design of digital systems. The appearance of families of homogeneous circuits practically eliminates all adaptation problems, and makes realization very easy.

(5) Programming Intelligence

Programming digital system gives them a sharp superiority over analogue ones, because they be made intelligent. This allows, for example, a choice between several strategies, according to circumstances. This facility can be used to insure safety of function, and to achieve tight control of the system.

In the first case, thresholds are defined for some variables, which may encounter dangerous variations. The apparatus analyses and processes the fault. In the second case, the system takes into account the encountered defaults, and changes the strategy of command in order to maintain the best possible function with regard to system evolution.

Programming also brings facilities for introducing constant factors, test conditions, iteration, etc. Thus it is possible to solve problems by man-machine dialogue, which is impossible by analogue means. The dialogue is realized with the aid of centralized command of programmed systems, which appears more and more efficient in process supervision of manufacturing industries.

9.5 CONCLUSION

Measurement has permitted refinement of human knowledge, and no longer lies in its traditional quarters, in the laboratory. It is increasingly spreading into the industrial domain where it plays an active role in production, ensuring, among others, quality control and accuracy of production such as in cement plants, refineries and electric plants, supervision of the proper functioning of processes in machine, automobile or aircraft test sites as well as in production plant etc., the analysis of diverse parameters, in order to optimize the functioning of an industrial facility (oil, gas, electric power transportation systems, etc.), regulation, optimization or safety enhancement in the functioning of a system. Its evolution is characterized by an ever-increasing number of parameters to be supervised, and ever more severe requirements for quality of measurement.

Progress in electronics and data processing have considerably contributed to this evolution. They lead to increasingly rational and systematic organization of the countless measurements that are now required in industry. These measurements include very diversified physical quantities (temperature, pressure, flowrate, mechanical strain, etc.) which may be spread over locations often remote from the measuring units. With the exception of data acquisition, all other operations are centralized in a station.

In cases of geographically dispersed units, a first centralization can be performed at the level of satellite stations, which then amplify the primary signals to a level sufficient for transmission. The measurement station must then decode, sort out and divert the signals to their end destination. In case of sampling measurements, the sampling itself must be programmed at the station. The signals are then digitized and processed. For safety reasons, supervision, regulation and control functions are often duplicated, and assumed by distinct computers.

The design of efficient measurement systems has until now been time-consuming and expensive. Realization of the measurement and supervision systems of some recent nuclear plants has taken many years. Despite this, choosing the elements of a system (the kind of sensors, the measurement method, mainframe or minicomputer, quantities to be supervised, number of measuring operations, processing method, etc.) still remains empirical. So it may be hoped that the next stage of evolution in this domain will be the possibility of rationalization of the choice, founded on rigorous analysis of function and performance of components, and their optimal association, with a view to making these measurement systems more efficient and economic.

Appendix
MEASUREMENT AND SYSTEMS OF UNITS OF MEASUREMENT

A1 Definitions

Measurement generally involves using an instrument as a physical means of determining a quantity or variable. The *electronic* instrument, as its name implies, is based on electrical and electronic principles for its measurement function.

Measurement work employs a number of terms which should be defined here.

— *Accuracy* is the closeness with which an instrument reading approaches the true value of the variable being measured.
— *Precision* specifies the degree to which successive measurements of the same true value differ from one another.
— *Sensitivity* is the response of the instrument to a unit change of input signal.
— *Resolution* is the smallest change in measured value to which the instrument will respond.
— *Error* is deviation from the true value of the measured variable.

No measurement can be made with perfect accuracy, but it is possible to find out what the accuracy actually is and how different errors have entered into the measurement.

A2 Systems of Units of Measurement

To perform calculations with physical quantities, we must define them both in *kind* and *magnitude*.

In physical measurement two kinds of units are used: *fundamental units* and *derived units*.

In the SI system (Système International d'Unités) six basic units whose symbols are listed in Table A1.1, are used.

Table A1.1 Basic SI quantities, units and symbols

Quantity	Unit	Symbol
Length	metre	m
Mass	kilogram	kg
Time	second	s
Electric current	ampere	A
Thermodynamic Temperature	kelvin	K
Luminous intensity	candela	Cd

Table A1.2 Fundamental and derived units

Quantity	Symbol	Dimension	Unit	Unit symbol
Fundamental				
Length	l	L	metre	m
Mass	m	M	kilogram	kg
Time	t	T	second	s
Electric current	I	I	ampere	A
Thermodynamic temperature	T	Θ	kelvin	K
Luminous intensity			candela	cd
Plane angle	α, β, γ	$[L]^0$	radian	rad
Solid angle	Ω	$[L^2]^0$	steradian	sr
Derived				
Area	A	L^2	square metre	m^2
Volume	V	L^3	cubic metre	m^3
Frequency	f	T^{-1}	hertz	Hz (1/s)
Density	ρ	$L^{-3}M$	kilogram per cubic metre	kg/m^3
Velocity	v	LT^{-1}	metre per second	m/s
Angular velocity	ω	$[L]^0 T^{-1}$	radian per second	rad/s
Acceleration	a	LT^{-2}	metre per second squared	m/s^2
Angular acceleration	α	$[L]^0 T^{-2}$	radian per second squared	rad/s^2
Force	F	LMT^{-2}	newton	$N\ (kg\ m/s^2)$
Pressure, stress	p	$L^{-1}MT^{-2}$	newton per square metre	N/m^2
Work, energy	W	L^2MT^{-2}	joule	$J\ (N\ m)$
Power	P	L^2MT^{-3}	watt	W (J/s)
Quantity of electricity	Q	TI	coulomb	C (A s)
Potential difference, electromotive force	V	$L^2MT^{-3}I^{-1}$	volt	V (W/A)
Electric fieldstrength	E, ε	$LMT^{-3}I^{-1}$	volt per metre	V/m
Electric resistance	R	$L^2MT^{-3}I^{-2}$	ohm	Ω (V/A)
Electric capacitance	C	$L^{-2}M^{-1}T^4I^2$	farad	F (A s/V)
Magnetic flux	Φ	$L^2MT^{-2}I^{-1}$	weber	Wb (v s)
Magnetic field strength	H	$L^{-1}I$	ampere per metre	A/m
Magnetic flux density	B	$MT^{-2}I^{-1}$	tesla	$T\ (Wb/m^2)$
Inductance	L	$L^2MT^{-2}I^2$	henry	H (V s/A)
Magnetomotive force	U	I	ampere	A
Luminous flux			lumen	lm (cd sr)
Luminance			candela per square metre	cd/m^2
Illumination			lux	$lx\ (lm/m^2)$

The SI system is replacing all other systems in the metric countries. The derived units are expressed in terms of these six basic units by defining equations.

Examples

$$\text{Velocity } v = \frac{\text{Length L}}{\text{Time T}} \rightarrow v = LT^{-1} \text{ m/s}$$

Quantity of electricity Q = electric current $I \times$ time T
$Q = IT$ A s

Table A1.2 gives a complete listing of the fundamental quantities and the derived units in the SI which are recommended by the *General Conference of Weights and Measures*. The Eleventh General Conference designated *radian* and *steradian* as supplementary units.

The Imperial system of units uses the *foot* (ft), the *pound-mass* (lb) and the *second* (s) as the three fundamental units of length, mass and time, respectively.

The *inch* has been fixed at 25.4 mm.

The measure for the pound (lb) has been fixed at 0.45359237 kg. These

Table A1.3 Conversion from Imperial to SI units

	Imperial unit	Symbol	Metric equivalent	Reciprocal
Length	1 foot	ft	30.48 cm	0.0328084
	1 inch	in.	25.4 mm	0.0383701
Area	1 square foot	ft^2	9.29030 × 10^2 cm^2	0.0107639 × 10^{-2}
	1 square inch	in^2	6.4516 × 10^2 mm^2	0.155000 × 10^{-2}
Volume	1 cubic foot	ft^3	0.0283168 m^3	35.3147
Mass	1 pound (avdp)	lb	0.45359237 kg	2.20462
Density	1 pound per cubic foot	lb/ft^3	16.0185 kg/m^3	0.062428
Velocity	1 foot per second	ft/s	0.3048 m/s	3.28084
Force	1 poundal	pdl	0.138255 N	7.23301
Work, energy	1 foot-poundal	ft pdl	0.0421401 J	23.7304
Power	1 horsepower	hp	745.7 W	0.00134102
Temperature	degree F	°F	5(t − 32)/9°C	—

two figures allow all units in the Imperial system to be converted into SI units. Table A1.3 lists some of the common conversion factors for Imperial into SI units.

A3 Standards of Measurement

A standard of measurement is a physical representation of a unit of measurement. For example, the kilogram, which is the fundamental unit of mass in the SI, is realized by the mass of the *International Prototype Kilogram* consisting of a platinum–iridium alloy cylinder preserved at the International Bureau of Weights and Measures at Sèvres (France).

Prototype standards have several defects. They can change and can be damaged. The best standards appear to be atomic standards, because all atoms of given isotopes of a given element are identical and invariant in their properties. In recent years two of the fundamental units, the second and the metre, have been defined as follows:

metre (m): The metre is the length equal to 1 650 763.73 wavelengths in vacuum of the radiation corresponding to the transition between the level $2_{p_{10}}$ and 5_{d_5} of the krypton-86 atom.

second (s): The second is the duration of 9 192 631 770 periods of the radiation corresponding to the transition between the two hyperfine levels of the ground state of the caesium-133 atom.

This conversion to atomic standards has been made possible by the development of electronic measuring systems. Only the kilogram involves a prototype standard. The kelvin and candela, while not exactly atomic standards, are defined as follows:

kelvin (K): The kelvin, the unit of thermodynamic temperature, is the fraction 1/273.16 of the thermodynamic temperature of the triple point of water.

candela (cd): The candela is the luminuous intensity, in the perpendicular direction of a surface of 1/600 000 m² of a black body at the temperature of freezing platinum under a pressure of 101 325 newtons/m².

These two units are less basic than the others, but are included in the basic set for convenience.

The connection between the mechanical and electrical standards is established by the ampere base unit, which is defined as follows:

ampere (A): The ampere is that constant current which, if maintained in two straight parallel conductors of infinite length and negligible circular cross section and placed 1 m apart in vacuum, would produce between the conductors a force equal to 2×10^{-7} newton/m of length.

The variation of inductance L with displacement x between two coils carrying current i, one fixed and the other on one arm of a balance, allows very accurate determination of the magnetic force F:

$$F = \tfrac{1}{2} i^2 \frac{dL}{dx}.$$

Having determined the ampere, the remaining electrical units may be given the following definitions:

volt (V): The volt is the difference of potential between two points of a conducting wire carrying a current of 1 A, when the power dissipated between these points is equal to 1 W.

ohm (Ω): The ohm is the resistance between two points of a conductor when a difference of potential of 1 V, applied between these points, produces in this conductor a current of 1 A.

coulomb (C): The coulomb is the quantity of electricity transported in 1 s by a current of 1 A.

farad (F): The farad is the capacitance of a capacitor between the plates of which there appears a difference of potential of 1 V when it is charged by a quantity of electricity equal to 1 C.

henry (H): The henry is the inductance of a closed circuit in which an e.m.f. of 1 V is produced when the current in the circuit varies at a rate of 1 A/s.

weber (Wb): The weber is the magnetix flux which, linking a circuit of one turn, produces in it an e.m.f. of 1 V as the flux is reduced to zero at a rate in 1 s.

tesla (T): The tesla is a flux density of 1 Wb/m^2.

Prototype standards remain the basis for most working standards of measurement. Thus standard electrochemical cells of various types and Zener diodes are used as voltage standards subject to occasional absolute calibration. In this appendix it is not possible to examine the whole hierarchy of standards. The purpose has been simply to remind the reader of some idea of the importance of precisely defined universal units in measurement which are used throughout this book.

No man can reveal to you aught but that which already lies half asleep in the dawning of your knowledge.
The teacher who walks in the shadow of the temple, among his followers, gives not of his wisdom but rather of his faith and his lovingness.

Kahlil Gibran
(The Prophet)

BIBLIOGRAPHY

1. *'Integrated Circuit Converters and Data Acquisition System'*, Analog Devices Inc., Norwood, Mass. 02062 USA, 1986.
2. *'Ultrasonic cross-correlation flowmeters'*, J. Coulthard, *Ultrasonics*, March 1973.
3. 'Recursive digital filters with linear-phase characteristics', P. A. Lynn, *The Computer Journal*, **15** (4), 1971.
4. 'Un filtre universel aux multiples applications,' A. Ripaux, *Electronique Applications*, n° 26, 1982.
5. *'Electronic Instrumentation and Measurement Techniques*, W. D. Cooper, Prentice-Hall, Englewood Cliffs, N.J. 07632, 1978.
6. *Acquisition et Traitement de Données*, R. Duperdu, Ecole supérieure d'électricité, 91190 Gif sur Yvette (France), 1983.
7. *'Perturbations'*, A. Malicet, Ecole supérieure d'électricité, 91190 Gif sur Yvette (France), 1982.
8. *'Guide d'Applications des Convertisseurs N/A C-MOS'*, Analog Devices S. A., 1982.
9. *'Advanced Electronic Circuits'*, U. Tietze, Ch. Schenk, Springer-Verlag, Berlin, Heidelberg, New York, 1978.
10. '2920 *Analog Signal Processor Design Handbook'*, Intel Corporation, 3065 Bowers Avenue, Santa-Clara, CA 95051 (408) 987-8080, August 1980.
11. *'TRW LSI Products'*, Agence en France: Radio-Equipements Antarès S. A., 9, rue Ernest Cognacq, 92301-Levallois-Perret Cedex, 1982.
12. *'Système d'acquisition de données analogiques et de stimulation'*, Yves Pichon et C. N. Truong, Electronique Industrielle, n° 43, December 1982.
13. *'MF 10 Universal Monolithic Dual Switched Capacitor Filter'*, National Semiconductor, April 1982.
14. *'Special Functions Databook'*, National Semiconductor, 1979.
15. *'The Optoelectronics Databook for Design Engineers'*, Texas Instruments, 379, avenue du Général de Gaulle 92-Clamart-France (Tel. 645 07.07), 1983.
16. *'Commande de four par microprocesseur'*, Mesures, Régulation, Automatisme, n° 1, January 1983.
17. *'Télémesures'*, Colette Pain, cours ESE (Ecole Supérieure d'Electricité, Gif-sur-Yvette) n° 2867, 1981.
18. *'HP-IB Installation and Theory of Operation Manual'*, Hewlett-Packard, April 1980.
19. *'Data Acquisition Modules, Data Translation'*, 4 Strathmore Rd, Natick MA 01760 (617) 6555300, USA, 1980.
20. *'Liaisons des Jauges aux Instruments'*, Vishay-Micromesures, 98, boulevard Gabriel-Péri, 92240 Malakoff, 1979.
21. *'Signal Analysis'*, Athanasios Papoulis, Polytechnic Institute of New York, McGraw-Hill.

22. '*Databook* 1981', Intersil, 217, bureaux de la Colline, Bât. D, 92213 Saint-Cloud Cedex, 1981.
23. '*General Catalog*', Burr-Brown, International Airport Industral. Park, P.O. Box 11300, Tucson, Arizona 85734, 1983.
24. '*Théorie de la Communication*', J. Dupraz, Edition Eyrolles, 1973.
25. '*Télémesures—Télécommandes dans les Véhicules Aéronautiques et Spatiaux*', P. Drouot, cours de session de perfectionnement E.S.E., 1979.
26. 'Différence de phase' Tran Tien Lang, *Technique de l'Ingénieur*, R 1045, 1980.
27. *Catalogue* 1977–1978 'Mesure de grandeurs mécaniques', Philips, 105, rue de Paris, 93002 Bobigny, 1977–1978.
28. *Phaselock Techniques*, F. M. Gardner, Wiley-Interscience, 605 Third Avenue, New-York, N.Y. 10016., 1979.
29. *Synthèse des réseaux Actifs*, Tran Tien Lang, Ecole speciale de mecanique et d'electrité-75006 Paris (France), 1975.
30. *Documentation Intersil*, Ref. 582.115, ICL 7115, 1982.
31. *IEEE Standard Digital Interface for Programmable Instrumentation*, ANSI/IEEE Std 488–1978, publié par IEEE, 345 East 47th Street, New York, N.Y. 10017, USA, 1978.
32. '*MOS Switched capacitor filter with reduced number of operational amplifiers*', *IEEE Journal of Solid State Circuits*, **SC14** (6), December 1979.
33. '*Improved circuits for the realization of switched capacitor filters*', K. Martin, *IEEE Journal of Solid State Circuits.*, **CAS 27** (4) April 1980.
34. MOS switched-capacitor filters, R. Brodersen, P. R. Gray, and D. A. Hodges, *IEEE Journal of Solid State Circuits*, **67** (1) January 1979.
35. '*Mise en oeuvre des procédés électroniques dans les techniques de mesures*', Tran Tien Lang, *Techniques de l'Ingenieur*, 21 rue Cassette, 75006 Paris (France), October 1984.
36. *Program for the Design of FIR Using the Remes Exchange algorithm*, J. Mclellan, Rice University, April 13, 1973.
37. *Electronic Measurements and Instrumentation*. B. M. Oliver and J. M. Gage, McGraw-Hill, Inter University Electronics Series, Vol 12.
38. 'Circuits fondamentaux de l'électronique analogique.' Tran Tien Lang, *Technique et Documentation*, Lavoisier, 11 rue Lavoisier, F75384 Paris Cedex 08, 1986.
39. 16/32 bit digital signal processor', *TMS* 320 10 *User's Guide*, Texas Instruments, 1985.
40. *Noise Reduction Techniques in Electronic Systems* H. W. Ott, Wiley-Interscience.
41. 'Traitement numérique des signaux', Murat Kunt, Edition Georgi., Ch. 1813 St Saphorin, 1980.
42. *Mesures et Instrumentation* Tran Tien Lang, Ecole supérieure d'électricité, 91190 Gif sur Yvette (France), 1985.
43. *Designing Microprocessor based Instrumentation*, J. J. Carr, Reston Publishing Company, Reston, Virginia,' A Prentice-Hall Company.
44. *Integrated Circuits*, data book SWT-503, Sprague, 1984.
45. *Méthodes et Techniques de Traitement du Signal et Applications aux Mesures Physiques*, J. Max, Masson, 1981.
46. *Linear and Conversion Products*, Precision Monolithics Incorporated Data Book (1984), 1500 Space Park Drive, Santa Clara, CA 95050 USA, 1985.
47. *Electronique Numérique*, Tran Tien Lang, Ecole speciale de mecanique et d'électricite, 4 rue Blaise Desgoffe, 75006 Paris (France), 1979.

INDEX

Absolute-value converter, 131
AC/DC converter, 130
Accuracy, 30, 38, 43, 73, 133, 134, 159, 167, 204, 220, 223, 230, 260, 287, 307
Acquisition, 255, 280, 281, 286
Acquisition time, 218, 257, 258, 259, 262, 267, 268
Active filter, 32, 39, 93
Active transducer, 21, 36
Address, 262, 264, 267, 269, 281, 295, 298
Amplitude modulation (AM), 117, 118
Amplifier, 37, 38, 46, 52, 55, 65, 77, 79, 81, 82, 84, 85, 88, 112
Analogue correlator, 140
Analogue multiplier, 42, 111, 120, 139, 151
Analogue switch, 87, 105, 165, 174, 175, 194
Analogue-to-digital converter (ADC), 14, 64, 193, 210, 211, 212, 220, 243
Aperture time, 255, 257, 259
Autocorrelation, 121, 139, 143, 144, 159, 166
Automatic measurement system, 294

Band-pass filter, 95, 100, 101, 117
Band-width, 9, 10, 11, 39, 79, 85, 144, 245
Base-emitter junction, 213
Bias current, 66, 167, 168, 178, 231
Binary decoder, 263
Bipolar transistor switch, 187
Bridge, 27, 28, 29, 30, 31, 40, 41, 56, 69, 115
Buffer, 158, 228
Bus, 291, 294, 295, 296, 297
Bypassing power supplies, 61

Capacitance, 49, 50, 202, 213, 214, 258, 260
Capacitive coupling, 49, 50
Capacitive transducer, 22, 33
Charge amplifier, 82, 83, 84
Chopper amplifier, 85
Clock, 210, 214, 229, 230, 245, 280, 281
Colpitts oscillator, 34
Common Mode Rejection Ratio (CMRR), 50, 59, 60, 61, 68, 69, 70, 71, 75, 230
Common Mode Noise Voltage, 57, 167
Comparators, 165, 166, 211, 213, 214, 222, 230
Complementary MOS (CMOS), 183, 184, 185, 199, 200
Computer, 276, 277, 280, 282, 284, 286, 287
Conditioner, 4, 5, 14, 27, 31, 32, 33, 36, 37
Conductive coupling, 53
Conversion rate, 207, 212
Conversion time, 212, 217, 218, 219, 220, 277
Converter, 123, 128, 130, 131, 132, 133, 135, 193, 210, 220, 221, 223, 224, 229, 233, 234
Consumption, 1, 302
Convolution, 92, 288, 292
Correlators, 139, 140, 141, 142, 144, 145, 161, 162
Current source, 36, 37, 196, 197, 201
Current-to-voltage converter, 38, 194

Data acquisition system, 16, 17, 193, 267, 270, 272
Data processing, 13
Decimal code, 194
Decoder, 158, 265

Decoupling filter, 63
Delay, 140, 169, 219, 257
Delay line, 107, 218, 219
Demodulation, 7, 118, 119, 137, 138, 139
Depletion, 178
Detector, 4, 5
Dielectric, 22, 23, 24, 178, 255
Differential amplifier, 55, 56, 57, 58, 209, 277
Differential input voltage, 57, 267
Differential multiplexer, 267
Differential transformer, 22, 31
Digital correlator, 143
Digital filter, 240, 243, 287, 288
Digital measuring system, 13
Digital multiplexer, 268, 269, 270
Digital multiplier accumulators, 287, 290
Digital phasemeter, 157
Digital processing, 286
Digital signal processor, 290, 292
Digital storage oscilloscope, 242, 244
Digital-to-analogue converter (DAC), 15, 91, 193, 200, 204, 205, 208, 209
Diode switch, 186
Drift, 75, 85, 86, 87, 207, 303
Drivers, 227, 228, 264
Dual slope converter, 229, 275

Effect
  electromagnetic, 34
  Hall, 34, 46, 47
  photoelectric, 34
  photovoltaic, 34
  piezoelectric, 34
  thermoelectric, 34
Emitter to base junction, 125
Enhancement mode MOSFET, 178, 179, 183
Equivalent circuit, 70, 200, 201
Errors, 122, 124, 129, 136, 167, 188, 189, 200, 204, 205, 231, 235, 240, 256, 307

Feedback, 40, 41, 203, 260
Feedback loop, 73, 78
Feedthrough, 190, 202, 256
Field Effect Transistor, 38, 67, 87, 174, 175, 177, 178, 231
Filter, 86, 93, 94, 95, 108, 109, 115, 274
Flash converter, 213, 273
Flow meter, 159, 162
FM, 8, 10

Fourier Transform, 6, 92, 121, 249, 251, 288, 291, 292
Frequency, 33, 50, 51, 63, 83, 85, 138, 139, 145, 151, 190
  cut-off frequency, 93, 97, 98
  free-running frequency, 146, 147
Frequency multiplexing, 6
Frequency response, 67, 252, 288
Function, 96, 97, 98, 103, 104, 106, 108, 114

Gate, 181, 182, 183
Gauge, 2, 3, 5, 23, 25, 26
General purpose interface Bus (GPIB), 295
Generalized impedance converters, 95, 96
Glitches, 206, 280
Ground, 51, 53, 63, 105
Grounding connection, 53, 55
Guard, 71, 72, 82, 267

Hall effect, 35, 46, 47, 111
Harley oscillator, 34
Hay Bridge, 31
High-pass filter, 94, 97, 98, 100
Hysteresis comparators, 170
Hysteresis error, 237

Inductive coupling, 51, 56
Inductive transducer, 22, 34
Input bias current, 75
Instrument amplifier, 5, 15, 29, 33, 46, 64, 65, 72, 74, 266
Interface, 4, 5, 91, 182, 183, 190, 243, 276, 284, 285, 294, 295, 296, 298, 299
Interfacing, 5, 21, 32, 36, 226
Interference, 3, 48, 49, 53
Integrator, 119, 120, 121, 224, 234, 261
International System of Unit (SI), 308
Isolation amplifier, 29, 33, 76, 77, 78, 79, 81
  optically-coupled isolation, 76, 77
  transformer-coupled isolation, 76, 77
Isolation Mode Rejection, 80

Junction FET, 175, 176, 177, 178, 180, 189

Ladder, 196, 197, 198
Ladder network, 195, 209
Laplace Transform, 83, 103, 107, 154

317

Leakage current, 77, 78, 79, 80, 178, 179, 180, 189, 259, 263
Leakage impedance, 71, 81
Least significant bit (LSB), 194, 199, 201, 203, 212, 216
Linearizing, 39, 115
Load, 62, 73, 90
Locking frequency, 156
Logarithmic converter, 42, 123, 124, 125, 127, 134
Loop filter, 147, 153
Loop stability, 154
Low-pass filter, 94, 95, 97, 98, 100, 102, 149, 152, 153, 160

Magnetic field, 47, 51
Maxwell Bridge, 31
Measuring system, 3, 4, 7, 13, 48, 164
Memory, 262, 263, 298
Memory address, 263, 291
Metal Oxide Semiconductor (MOS), 178, 179, 180, 181, 182, 183, 189, 191
Metal strain gauge, 29
Microprocessor, 14, 15, 91, 209, 210, 215, 219, 286, 288, 290, 299
Modulation, 4, 5, 7, 8, 10, 13, 117, 118
Monolithic transducer, 44, 45, 46
Monolithic correlator, 144, 145
Monotonicity, 205, 206
Most Significant Bit (MSB), 195, 196, 198, 199, 200, 215
Multiplexer, 17, 247, 263, 264, 265, 266, 267, 284
Multiplier, 112, 114, 117, 134, 142, 149, 157

n-channel MOS, 178, 179, 183
Noise, 48, 50, 51, 121, 134, 238, 300, 303
Non-linearity, 29, 41, 42, 43, 67, 205, 207, 225, 230

Offset voltage, 66, 168, 175, 178, 260, 275
Open collector output, 169, 295
Operational amplifiers, 37, 38, 65, 67, 72, 75, 78, 93, 94, 95, 112, 129, 131
Oscillator, 33, 34, 39, 85
Oscilloscope, 210, 242, 244

Passive transducers, 21, 26, 28
p-channel MOS, 178, 179, 180, 182, 183, 184

Peak-value converter, 135, 136, 137
Permeability, 1, 21, 24
Permittivity, 1, 21, 23
Phase comparator, 146, 150
Phase difference, 119, 148, 157, 242
Phase error detector, 146, 151, 152, 154, 161
Phase locked loop (PLL), 145, 147, 148, 156, 157
Phase measurement, 119
Photoelectric effect, 35, 36
Photovoltaic effect, 35, 36
Piezoelectric effect, 35, 36
Piezoresistive accelerometer, 4
Pinchoff voltage, 137, 175
Potentiometer, 86, 228
Power supply, 53, 54, 61, 79, 208
Pulse Amplitude Modulation (PAM), 11
Pulse Code Modulation (PCM), 13
Pulse Duration Modulation (PDM), 13
Pulse Position Modulation (PPM), 13
Precision, 30, 87, 122, 307
Processing, 41, 114, 286, 290, 294, 304
Programmable filter, 208
Programmable gain amplifier, 17, 75, 88, 274, 282, 284

Quad slope converter, 234, 275
Quantization, 237, 239, 300, 302
Quantizing, 248
Quantum, 248
Quartz oscilator, 222

$R-2R$ network, 197, 198, 199, 209
Ramp voltage, 210, 221
Ranch Cell Filter, 96
Read Only Memory (ROM), 284, 292
Real-time correlator, 140
Recorder, 91
Rectifier, 131, 132, 137, 152
Reference voltage, 194, 195, 196, 200, 213, 216, 235, 291
Register, 91, 142, 214, 218, 288
Reliability, 2
Resistivity, 21, 22, 23
Resolution, 30, 75, 204, 207, 212, 214, 253, 254, 307
Response time, 167, 168, 169
Rise time, 245, 263
Root mean square (RMS), 132, 135, 238
Row correlators, 141

Sample-and-hold, 14, 212, 216, 217, 218, 241, 247, 251, 255, 259, 261, 262, 277
Sampling frequency, 243, 245, 247, 250, 254, 274, 280
Sampling rate, 244, 247, 252, 253, 272, 274
Sampling theorem, 241, 249
Saw-tooth generator, 222, 223, 244
Scale factor, 127, 133
Schmitt trigger, 170, 171, 172
Sensitivity, 1, 24, 26, 27, 29, 31, 45, 47, 86, 93, 96, 120, 146, 150, 157, 160, 307
Series Mode Noise Rejection, 238
Settling time, 39, 206, 245
Shield, 50, 52, 61, 267
Shielded cables, 60
Signal conditioner, 3, 13, 14, 27, 31, 33, 76
Signal processing, 114, 294
Signal-to-noise ratio, 9, 122, 123, 144
Single-ended input, 263
Single-slope converter, 220, 221
Slew rate, 39, 165
Spectrum, 6, 8, 92, 93, 249, 250, 251, 252, 273, 292, 293, 294
Spectrum analysis, 291
Stability, 125, 225, 229, 230
Standards of measurement, 309, 311
Strain gauge, 5, 23, 25, 26, 227
Stray capacitance, 71, 105, 196
Substrate, 180, 202
Successive-approximation analogue-to-digital converter, 214, 275
Switch, 88, 183, 184, 185, 186, 187, 247
Switched capacitor filters, 98, 109, 110
Switched capacitor integrator, 104
Switching time, 192
Synchronous demodulator, 32, 119, 137, 138, 139
Synchronous detection, 119, 137
System of units, 307
 of measurements, 307

Temperature compensation, 125
Thermal effect conversion, 133
Thermistor, 40
Thermocouple, 15, 68, 227
Thermoelectric effect, 35
Thermoelectricity, 1
Threefold slope converter, 233, 234
Threshold voltage, 233, 234
Time multiplexing, 11
Track-and-hold amplifier, 213
Transducer, 21, 29, 34, 36, 37, 38, 39, 40, 41, 226, 227
 capacitive, 22, 33, 34
 inductive, 22
 modulating, 21
 piezoelectric, 82
 pressure, 24, 31, 45
 resistive, 22, 31, 33
 temperature, 44, 45
Transient, 191, 192, 254
Transient signal, 254, 262
Transistor, 34, 124, 127, 133, 167, 197, 209
Transmission, 8, 9, 11, 62, 123
Twisted conductors, 52

Units of measurements, 307, 308, 309, 310, 311

Voltage comparators, 165, 167
Voltage controlled oscillator (VCO), 111, 146, 147, 153, 154, 157
Voltage divider circuit, 27
Voltage source transducer, 37
Voltage-to-current converter, 90, 91
Voltage-to-frequency converter, 223, 224, 225, 226

Window comparator, 169

Z-transform, 103, 104, 107, 241, 288
Zero crossing detector, 119